I0473961

ALGEBRA EXAMPLES

TRIGONOMETRY 1

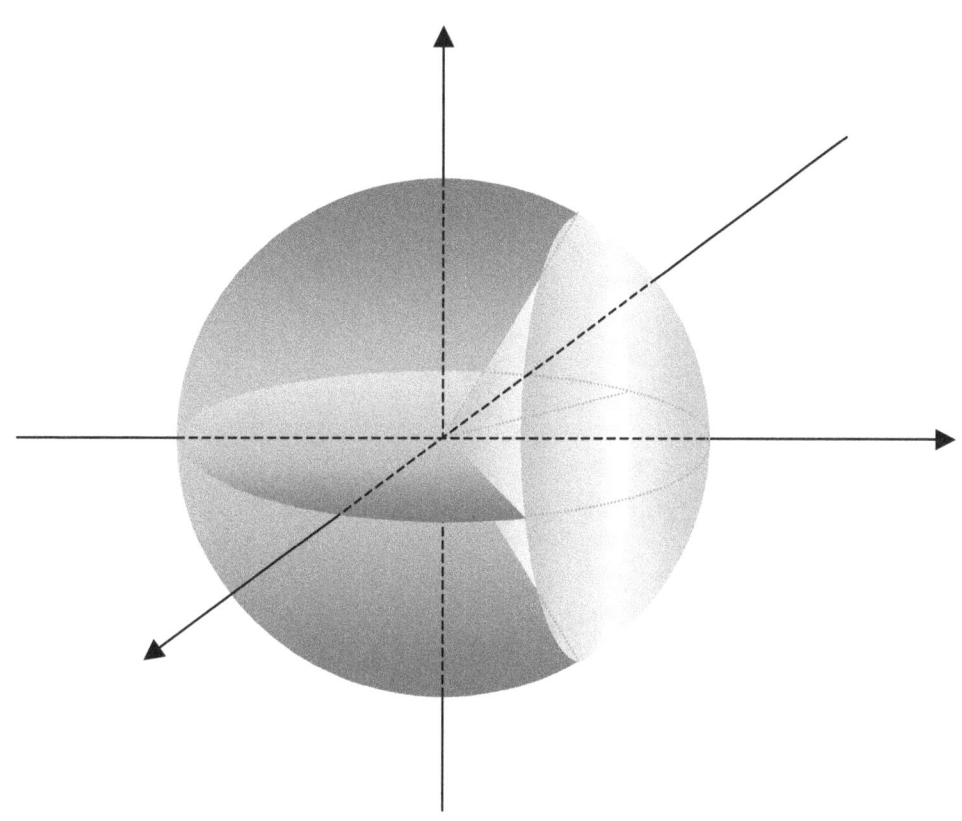

Seong R. KIM

Dear students:

Students need the best teacher, so you need examples, because examples are the best teacher. All the examples here are fully worked, and explain **how** the basic and essential tools in math are made, together with **what** they are, **how** they work, and **how** to work with them. Such tools include numbers, formulas, identities, equations, laws, etc.

Examples here begin with easy ones, of course. Covering every meter and yard properly, we can cover thousands of miles and kilometers. And it is particularly the case in math.

Of those examples therefore, some might even look too easy for you. It's not that easy though, to come up with those examples. Anyways, the bigger and the taller the tree, the deeper and the stronger the root.

Doing math, we work with ideas and run ideas, because every thing in math is an idea. A number is an idea, for instance, and the same is true for a line or circle, too. And putting ideas together, we build another, which becomes the base or an element of another, and each is connected. And that's the way your math grows. So you get to build a circuit, and sometimes, need to fill the gap or repair the circuit so that you get the sense of it.

So your calculation runs properly, and you get the problem solved.

The examples have been made and arranged so that they get tougher (or sometimes easier for some reason) as you proceed with them. In particular, similar examples with some variations are strategically repeated so that you can get the ideas or the tools tricky or complicated, and can get them mastered.

This book is however, nothing but a bunch of examples until you get it powered.   How then, to get it powered, and make it run and work for you?

Just read it, and then, do each example in writing. And it is important to note that you do it in **your** writing. Just watching someone doing it, you just only feel that you can do it. If you do it, you can do it, but if you don't, we can hardly. It's a cliché, of course, but is always true that knowing is one thing and doing is another.

I've been helping students grow, take care of, and run their own math. The area covers algebra and geometry for high school or college students, and is especially for equations (for unknowns or curves), functions, and their graphs, which are the basic elements in calculus, which's been the core of my interest from my early age in high school.

Of my students, some are quite poor in math, and thus, are afraid of or hate math, some require special education because of exceptional intelligence, some are smart enough, some are naïve and diligent, some are clever but lazy, and most behave in general. All the students are badly after though, one thing in common: a strong and secure math skill. It is of course, the prime objective of my work, and I'm always happy to and eager to help them achieve it. The problem was however, that many of them wanted it to be purchased. And the question is, can we buy it?

We can buy the means, of course. And a solid math skill is feasible, too. We know however, we can't buy love, and the same is true for the math skill, too. It's not what we can buy or sell, and not what we can give or take. It is however, what we can grow, and need to grow. Your math grows as much as you grow and take care of it. So does mine.

What math then, do students most often do or use in high schools or colleges?

It is algebra and geometry. What algebra though?

Elementary algebra, of course
Doing the algebra, we work with numbers (many in kinds), constants, variables, ratios, rates, expressions, equations, inequalities, functions, identities, formulas, laws, etc., together with signs and symbols. And if we want to do algebra properly, we want to know their natures and how they mingle with each other.

So studying math ideas or tools, you want to know **what** they are, **how** they work, and **how** to work with them or **what** to do with them.   What then, about the geometry?

Basically, the geometry has much to do with shapes, positions, and angles. The shapes begin with triangles and circles, and move on to rectangles, squares, parallelograms or rhombuses, trapezoids, tetragons, other polygons, polyhedrons, etc.

Doing the geometry, too, though, we need to do the algebra stated above. So it is analytic geometry, often called coordinate geometry, too. And doing it, we can specify positions using coordinates. So in the geometry, basically, we work with graphs. Putting a math idea in a graph, we can not only effectively think about it but actually see it, too, and therefore, can efficiently work with it. What idea then, is it?

The idea begins with a point, line, parabola, circle, ellipse, and hyperbola, called a conic section or basic curve, and then, moves on to other curves, planes, surfaces, volumes, and other objects in various dimensional spaces, together with vectors.

And using an angle, we can specify an amount of turn or change in direction.

So learning, using, or applying those ideas or math tools, we get to solve problems.

And this book can help. It can help learn them, and use them so that you can navigate to find solutions to problems. And in particular, it can help come up with answers to those **what**s and **how**s stated above. So it can help you grow and run your own math, and thus, can help achieve your solid math skill.

It is however, not a magic book giving you a math skill of high caliber overnight. And it can have many mistakes, too. There is no magic, and math is full of facts and ideas. And it is after all, not me and not your teacher but you who put together some of those facts and ideas, and understand it. Putting facts and ideas together, understanding it, and taking care of what you have learned, you grow your math. And this book can help.

This is a book of examples designed to help you grow your math, and assumes that you are a real beginner. This book requires though, time and effort, the amount of which need to be substantial, too, but will be worth it. That's because you want a substantial achievement, and will get it. And probably, you will get to see this book helping you get there much faster than expected. And then, you will get to see the way math runs.

In math, everything is an idea. So is a problem. And solving it, we put it many different ways. For instance, while expanding or reducing it, or modifying or converting it, we keep searching for the solution, approaching the solution, and eventually, can get there. So don't look for the solution outside the problem. The solution is inside the problem if the problem is properly made.

If it is not, no solution is the solution. And in fact, it is often the case a problem itself is the solution. We can put a problem in many different ways, and eventually, can end up with the solution.   How come then, is the solution no other than the problem?

For instance, the solution to $3232 \div 101$ is 32.   And we can put it this way:

$$3232 \div 101 = \frac{3232}{101} = \frac{32 \times 101}{101} = \frac{32}{1} = 32 \implies 3232 \div 101 = 32.$$

And we can get this, too: $32 \implies 3232 \div 101.$ How?

$$32 = \frac{32}{1} = \frac{32 \times 101}{101} = \frac{3232}{101} = 3232/101 = 3232 \div 101.$$ Too easy?

For another instance, the solution to $ax^2 + bx + c = 0$ is: $x = \frac{-b \pm \sqrt{b^2 - 4ac}}{2a}$, which is called the quadratic formula.   How come then, is the solution no other than the problem?

We can put it this way:

$$x = \frac{-b \pm \sqrt{b^2-4ac}}{2a} \implies 2ax = -b \pm \sqrt{b^2 - 4ac} \implies 2ax + b = \pm\sqrt{b^2 - 4ac}$$

$$\implies (2ax + b)^2 = b^2 - 4ac \implies 4a^2x^2 + 4abx + b^2 = b^2 - 4ac$$

$$\implies 4a^2x^2 + 4abx = -4ac \implies ax^2 + bx = -c \implies ax^2 + bx + c = 0.$$

And we can get this, too: $ax^2 + bx + c = 0 \implies x = \frac{-b \pm \sqrt{b^2-4ac}}{2a}.$ How?

$$ax^2 + bx + c = a(x^2 + \tfrac{b}{a}x) + c = a(x^2 + \tfrac{b}{a}x + \tfrac{b^2}{4a^2} - \tfrac{b^2}{4a^2}) + c = a(x^2 + \tfrac{b}{a}x + \tfrac{b^2}{4a^2}) - \tfrac{b^2}{4a} + c$$

$$= a(x + \tfrac{b}{2a})^2 - \tfrac{b^2-4ac}{4a} = 0 \implies a(x + \tfrac{b}{2a})^2 = \tfrac{b^2-4ac}{4a} \implies (x + \tfrac{b}{2a})^2 = \tfrac{b^2-4ac}{4a^2} \implies x + \tfrac{b}{2a} = \pm\sqrt{\tfrac{b^2-4ac}{4a^2}}$$

$$\implies x = -\tfrac{b}{2a} \pm \tfrac{\sqrt{b^2-4ac}}{2a} = \tfrac{-b \pm \sqrt{b^2-4ac}}{2a} \implies x = \tfrac{-b \pm \sqrt{b^2-4ac}}{2a}.$$

And we call the set of processes above, algebra.

So if a problem is well defined, that is, if it makes sense, we should be able to get it solved the way below:

A problem $\Rightarrow$ ... $\Rightarrow$ ... $\Rightarrow$ the solution, and thus: **the problem $\Rightarrow$ the solution**.

So solving a problem, we put it many different ways so that we can get to the solution.

And that's the way, math runs.

May your math run very well.

Seong R. Kim

B.S. Math. Michigan Tech. Univ.   M.S. Math. Rensselaer Polytechnic Institute

Notes:

This book is about a math idea called trigonometry.

Why trigonometry though?

That's primarily because we often get to work with angles not only doing geometry but doing algebra, too. Doing high school math or college math, we can hardly avoid or stay away from algebra. And doing geometry, too, we often get to do algebra on expressions with angles. And doing such algebra, we can say that we do trig-algebra.

And next, we need to do trigonometry if we have to work with vectors and many other objects that have to do with angles. Working with such objects, we often need to find objects called components, which have directions or angles. Finding such components, we want to use some tools in trigonometry. And the tools are called trigonometric-ratios, often just called trig-ratios, for short.

So doing trigonometry, we get to use trig-ratios called sines, cosines, etc., together with a bit bigger tools called trig-identities, and some rules or formulas. So in this book, you get to know what those tools are and what they are about. That is to say that you will get to know how those tools work, what you can do with them, and how to work with them.

Specifically, you will get to learn, for instance, how the trig-ratio called the sine is made, what it is about, and how to use it. More specifically, you will see why you have to multiply by the sine, and what you get multiplying by it.

And of course, you will get to learn those important tools called trig-identities and the formulas called the Sine Rule and the Cosine Rule. Besides, you will get to see and will be familiar with some special tools called trig-functions.

And you will get them all through examples, that is, those tools will get explained with examples fully worked and detailed. Also, following steps to the solution in each example, you will be more familiar with the tools and the math ideas.

And you will get to strengthen your skill of algebra. So doing problems as well as learning ideas in math, you can do better and faster so that your math can run not only properly but fast enough, too.

And all the basics, tools, and ideas are covered in three books as well as in one book. And the three are as follows:

ALGEBRA EXAMPLES TRIGONOMETRY 1, which is this book, and covers from the section Intro 1 to the section The Cosine Rule.

ALGEBRA EXAMPLES TRIGONOMETRY 2, which covers from Examples 1 in The Cosine Rule to the section Sine Functions.

ALGEBRA EXAMPLES TRIGONOMETRY 3, which covers from Examples in Sine Functions to Examples in Trig-Algebra.

And all the contents of the three books above are put in one book as follows:

ALGEBRA EXAMPLES TRIGONOMETRY, which covers thus, from the section Intro 1 to Examples in Trig-Algebra.

Contents

In TRIGONOMETRY 1

The Preview of the Contents

In TRIGONOMETRY 2

The Preview of the Contents

In TRIGONOMETRY 3

$$(x + y)^2 = x^2 + 2xy + y^2.$$

$$(x + y)^3 = x^3 + 3x^2y + 3xy^2 + y^3.$$

$$(x + y)(x - y) = x^2 - y^2.$$

$$(x + y)(x^2 - xy + y^2) = x^3 + y^3.$$

$$(x^2 + xy + y^2)(x^2 - xy + y^2) = x^4 + x^2y^2 + y^4.$$

$$(x + a)(x + b) = x^2 + (a + b)x + ab.$$

$$(ax + b)(cx + d) = acx^2 + (ad + bc)x + bd.$$

$$(x + a)(x + b)(x + c) = x^3 + (a + b + c)x^2 + (ac + bc + ca)x + abc.$$

$$(a + b + c)^2 = a^2 + b^2 + c^2 + 2(ab + bc + ca).$$

$$(a + b + c)(a^2 + b^2 + c^2 - ab - bc - ca) = a^3 + b^3 + c^3 - 3abc.$$

Suppose both a and $b \neq 0$, and both m and n are integers. Then, we get:

0. $a^m a^n = a^{m+n}$ **1.** $a^m/a^n = \dfrac{a^m}{a^n} = a^{m-n}$ **2.** $(a^m)^n = a^{mn}$

3. $(ab)^n = a^n b^n$ **4.** $(a/b)^n = \left(\dfrac{a}{b}\right)^n = a^n/b^n = \dfrac{a^n}{b^n}$

Suppose both a and $b > 0$, and m and n both are integers nonzero. Then, we get:

0.1. $a^{\frac{1}{n}} b^{\frac{1}{n}} = (ab)^{\frac{1}{n}}$.   **1.1.** $\dfrac{a^{\frac{1}{n}}}{b^{\frac{1}{n}}} = \left(\dfrac{a}{b}\right)^{\frac{1}{n}}$.   **2.1.** $(a^{\frac{1}{n}})^m = (a^m)^{\frac{1}{n}}$.

3.1. $(a^{\frac{1}{n}})^{\frac{1}{m}} = a^{\frac{1}{mn}} = (a^{\frac{1}{m}})^{\frac{1}{n}}$.   **3.2.** $(a^{mp})^{\frac{1}{np}} = (a^m)^{\frac{1}{n}}$, where p is a nonzero integer.

1. Suppose M, N, and $b > 0$, but $b \neq 1$, and we have: $A = \log_b M$, and $B = \log_b N$. Then, we get: $A - B = \log_b M - \log_b N = \log_b \frac{M}{N}$.

2. Suppose that M and $b > 0$, but $b \neq 1$, and that we have: $E = \log_b M$. Then, we get: $PE = P \log_b M = \log_b M^P$.

3. Suppose that a, b, C, and $D > 0$, but a and $b \neq 1$, and that we have: $\log_a C = \log_b D$. Then, we get: $\log_a C = \log_b D = \log_{ab} CD$.

4. Suppose that a, b, C, and $D > 0$, but a and $b \neq 1$, and that we have: $\log_a C = \log_b D$. Then, we get: $\log_a C = \log_b D = \log_{\frac{a}{b}} \frac{C}{D} = \log_{\frac{b}{a}} \frac{D}{C}$.

5. $\log_b b = 1$, and $\log_b 1 = 0$.   **6.** $\log_b A = \dfrac{\log_c A}{\log_c b}$.

7. $\log_b A = \dfrac{1}{\log_A b}$.

Note:

The drawings or graphs in this book are not exact, and are approximate or conceptual ones.

$\in$	"$a \in B$" means that a belongs to B. "$p, q,$ and $r \in W$" means that $p, q,$ and r belong to W.
$\Rightarrow$	"$A \Rightarrow B$." means that A implies B.
$\equiv$	$A \equiv B$ means that A and B are identical to each other.
$\neq$	$A \neq B$ means that A is not equal to B.
$|A|$	The magnitude of A. For instance, $|-1| = |1| = 1$.
$\therefore$	Therefore
$\Leftrightarrow$	"$A \Leftrightarrow B$" means "If A then B." and "If B then A." We can read $A \Leftrightarrow B$ as "A if and only if B." In such a case, we can say that $A = B$.
Δx and Δy	Suppose that (x_1, y_1) and (x_2, y_2) are two points in the x-y plane. Then, we get either of the two below. $\Delta x = x_2 - x_1$, and $\Delta y = y_2 - y_1$. $\Delta x = x_1 - x_2$, and $\Delta y = y_1 - y_2$.

Distance Formula

Suppose that d is the distance between two points (x_1, y_1) and (x_2, y_2) in the x-y plane. Then, we get $d^2 = (\Delta x)^2 + (\Delta y)^2$.

0.0. Intro 1

To begin with, trigonometry is not an area we can easily get used to.
So explanations on some ideas, tools, or objects will be *repeated*. It will repeat though, a bit differently each time, along with more details and examples *added* or *modified*.

So what is trigonometry, and what is it about?

Trigonometry is probably a composite word made of 'trigon' and 'geometry'. A trigon is an ancient triangular harp, and thus, is an old word for a triangle. It is in fact, a polygon made of three line segments. So trigonometry can be called *triangle geometry*.
It is however, not only about triangles but about circles, too.   Why circles though?

Trigonometry has much to do with angles. And if there were no circle, there would be no angle. So learning trigonometry, you need to know angles and circles. Not just circles though. You want to know the nature of circles, too, and be familiar with the nature.

And there would be no angle or direction, curved line or curved surface if a circle did not exist (in your knowledge). And in turn, you could not make any machine. So you cannot make a machine without (the idea of) a circle.   How come?

We may want to begin with an idea called an angle.

Suppose that two rays diverge from a point. Suppose also, that one of the two rays turns about the point towards the other until both rays cover each other.

Then, the amount of turning is an angle.

Fig. 0

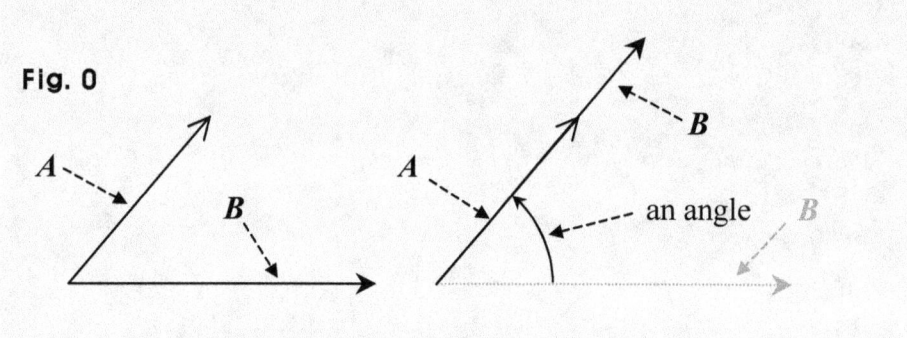

Suppose this time, the ray **B** turns about one of the points in the ray **B** itself until it gets parallel to the other ray **A**. Then, the amount of turning is an angle, too, which is the same as the angle mentioned above.

Fig. 1

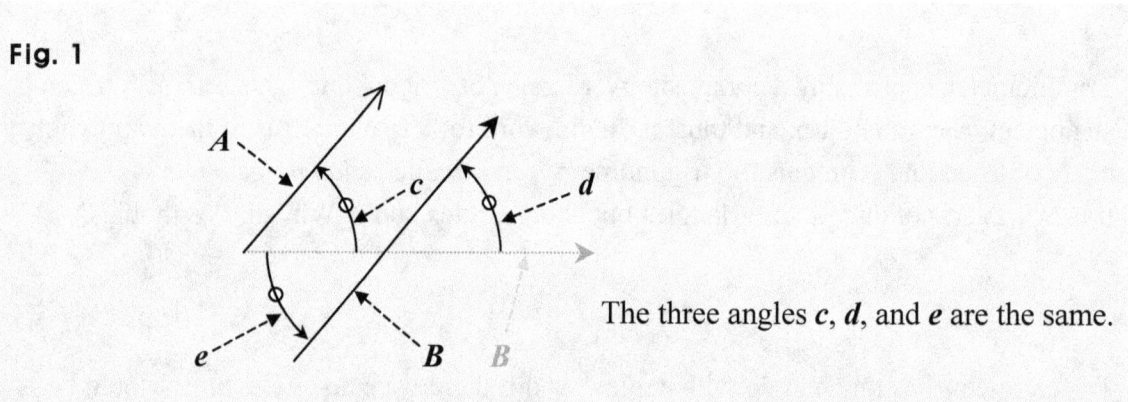

The three angles **c**, **d**, and **e** are the same.

Now, the ray **B** turned is parallel to the ray **A**, and has the same direction as that of the ray **A**. And we know the ray **B** before turning has a different direction.
What then, can we say about an angle?

We can say it is an amount of difference in direction, or an amount of change in direction. So in short, an angle can be called a change in direction, too.

And thus, we can use an angle specifying an amount of turning or a change in direction.

In short, *an angle* is *an amount of turning* or *a change in direction*.

A dictionary says an angle is the figure formed by two lines diverging from the point where the two lines meet, or is the figure formed by two planes diverging from the line where the two planes meet.

What figure then, do we normally use as the figure stated above?

We use as such a figure a part of a circle, which is called an *arc*.

And we can calculate an angle by means of an arc and a circle.

More precisely, we can get an amount of angle using a ratio of an arc length to the circumference of the circle the arc belongs to.

In short, an angle can be put *in terms of a ratio of an arc to the circle* that has the arc.

So for instance, assuming θ is an angle, k is such a ratio, and C is a constant, we can set: $\theta = Ck$.   What then, is the constant C?

It is $360°$.   So we get: $\theta = 360°k$.   Why is it $360°$ though?

Suppose a terminal ray, that is, a ray with a finite length is turning about the origin in the *x-y* plane as shown below.   Then, it keeps changing its direction, of course.

Fig. 2

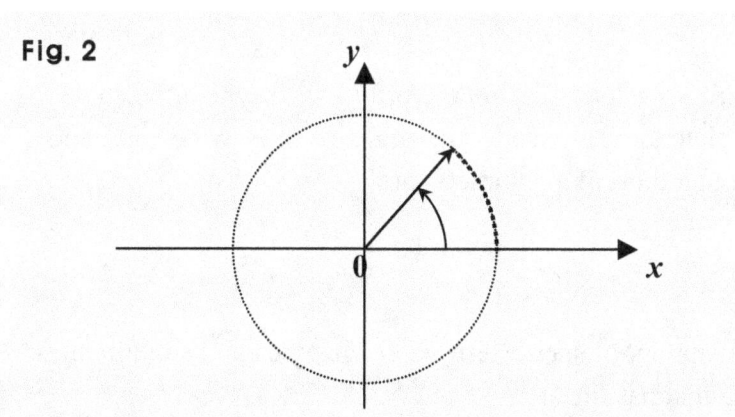

What do we mean by *directions* though?

NEWS are not the only directions, of course. We also have directions other than north, east, west, and south. We know north and west or east are perpendicular to each other, that is, at 90°, and in fact, we can precisely indicate a direction by means of an angle.

Let's now, get back to the figure above, where the terminal ray is turning.

Normally, it is assumed that the ray is *initially* on the **x**-axis on the right of the origin.

That is, before the ray starts turning, it is resting on the **x**-axis on the right of the origin.

And we say that *initially*, that is, when the ray is resting on the **x**-axis, its angle is **0°**.

Suppose now, the ray starts turning counterclockwise, about the origin, of course. Then, as the ray keeps turning, it keeps changing its direction, and its angle keeps changing.

And in this case, that is, when it's tuning ***counterclockwise***, the angle is said to be ***positive*** and *increasing*. And knowing the angle at a particular moment, we can specify the direction of the ray, at the particular moment, of course.

How then, do we get the angle?

If the ray makes a complete turn (or rotation) about the origin, we say that the angle made is 360°. And if two complete turns are made, the angle made is twice 360°, and thus, is 720°.   What then, about a third of a complete turn?

We can indicate by an arc, the amount of such a turn. And an angle can be put in terms of a ratio of an arc to the circle that has the arc.

Fig. 3

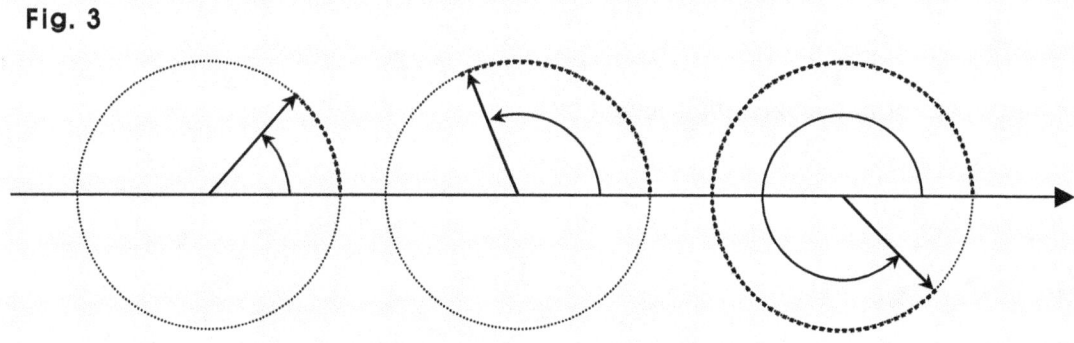

So if it makes a third of a complete turn, the arc is a third of the circle the arc belongs to.

And the ratio is a third to one, and thus, is simply a third.

So the angle made is a third of 360°, that is, 120°.

And thus, we can specify an angle by the product of 360° and the ratio of the arc to the circle the arc belongs to.

So for instance, assuming the length of an arc is 2, and 8 is the circumference of the circle where the arc belongs, we get: 2/8 of 360°, which is 1/4 of 360°, which is 90°.

So assuming θ is an angle, and k is such a ratio, we can set: $\theta = 360°k$.
And next, how come we cannot make any machine without a circle?

First of all, in a machine, something has to be able to turn. If nothing turns in a machine, the machine doesn't work.

And next, an amount of turning is an angle, which is from an arc, which is from a circle. So if no circle, then no turning, and thus, no machine.

Why does turning matter though? What does it have to do with a machine?

Normally, a machine repeats some same task, or makes something keep moving.

That's basically because turning happens in a machine.

How then, can turning make things repeat or keep moving?

Back to the ray turning.

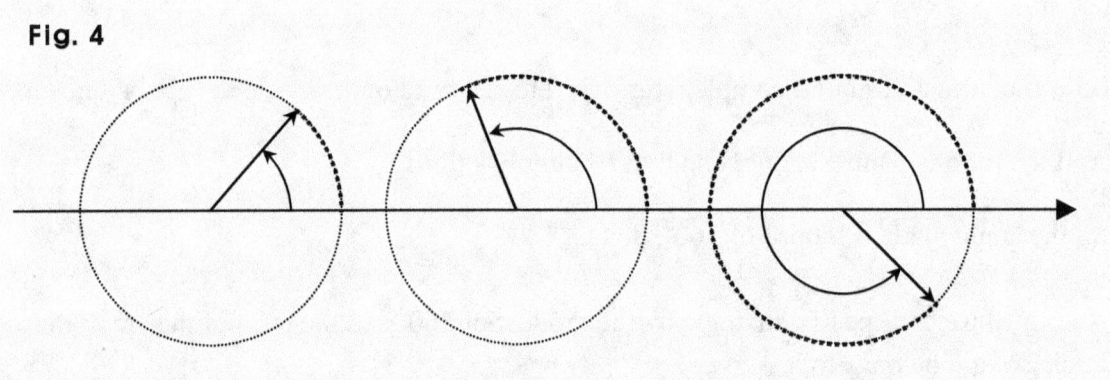

Fig. 4

As the ray keeps turning making complete turns over and over, the circular motion repeats. And the circular motion can get transferred to a linear motion.

Also, the circular motion repeating makes the linear motion continue or keep repeating.

For instance, as a wheel keeps turning, the wheel itself can be in a linear motion moving forward (or backward). So for instance, a bicycle can move forward or backward.

(And in fact, the same is true for a ball (sphere), too.)

How then, does a linear motion repeat?

Suppose, in the x-y plane where the ray is turning counterclockwise, we place two lamps the way below so that the shadow of the ray (the projection) is made on the x-axis.

Fig. 5

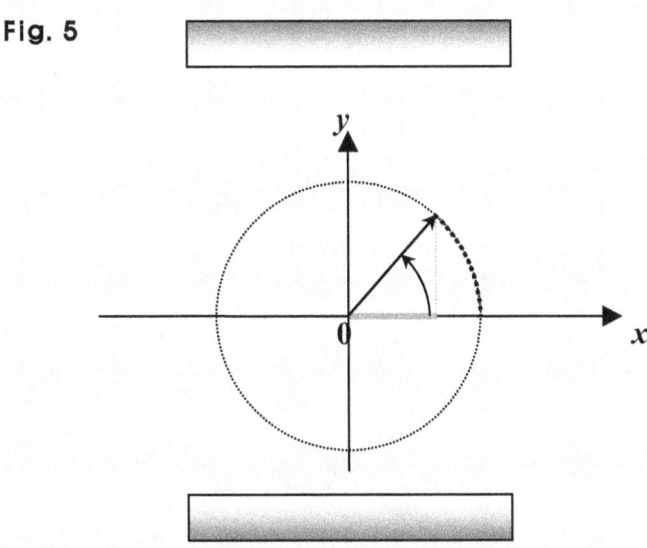

Then, as the ray keeps turning, we can see from above the *x*-axis, a linear motion where the shadow (projection) decreases, and then, increases, and also, we can see from below the *x*-axis, another linear motion where the shadow decreases, and then, increases. And also, we can see that those linear motions keep repeating as the ray keeps turning.

Suppose next, we place two lamps the way below so that the shadow (projection) is made on the *y*-axis.

Fig. 6

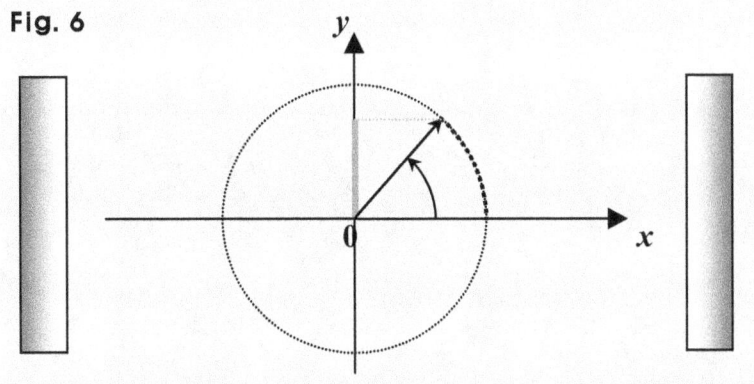

Then, as the ray keeps turning, we can see from the left of the *y*-axis, a linear motion where the shadow decreases, and then, increases, and also, we can see from the right of the *y*-axis, another linear motion where the shadow decreases, and then, increases. And also, we can see that such linear motions keep repeating as the ray keeps turning.

So what does such a linear motion have to do with trigonometry?

Trigonometry explains how the linear motion is related to the circular motion of the ray. And in the explanations, special ratios called trigonometric ratios are used, and that is what's covered in the next section, **Intro 2**, and the subsequent sections.

0.1. **Intro 2**

So what is trigonometry about?

Trigonometry relates angles and objects as lengths, areas, forces, etc. So working on objects in connection with angles, we run trigonometry. Running trigonometry, we use tools in trigonometry. And also, we often modify those tools or put them together to make another tool, too. In other words, we get to develop tools, too, because we need to solve problems, of course.

How then, do we know how modify the tools, or how to put them together?

Knowing how the tools are made and how they work, we can see what to do with them.

Then, we can run trigonometry nicely and powerfully, too. And thus, just as we do when learning other areas of mathematics, we want to know how the tools are made when learning trigonometry, too. And of course, in this book, you will get to learn how the tools are made and how they work, and how to modify or put them together, etc.

What tools though?

Beginning with the tools basic, we have three kinds of ratios, called *trigonometric ratios*, often just called *trig-ratios*, for short.

A trigonometric ratio is basically made of two sides in a right triangle. And we have three basic trig-ratios.   What then, are those three?

Getting a trigonometric ratio, we use two sides in a right triangle, and then, take the ratio between the two. We don't just pick two sides though. How then, do we choose the two?

Two particular sides make an important angle in a right triangle. And we may want to call the important angle the *governing angle*. That's because such an angle governs or determines the ratios.   What sides are the two though?

One is called the adjacent, and the other is called the hypotenuse. So in a right triangle, the *adjacent* and the *hypotenuse* make the *governing angle*.

• And in return, the *governing angle determines every trig-ratio* in the right triangle.

Depending on the way we look at the right triangle though, either of the two legs can be the adjacent. So using a wrong side as the adjacent, we get wrong ratios. Thus in short, the wrong adjacent makes wrong ratios.

So it is crucial to choose the adjacent correctly.   What side then, is the adjacent?

It is the leg *next to*, that is, *adjacent to* the governing angle. And we call the other leg the opposite, which faces the governing angle, and thus, is opposite of the governing angle.

So assuming the governing angle is *θ*, read as theta, we can get:

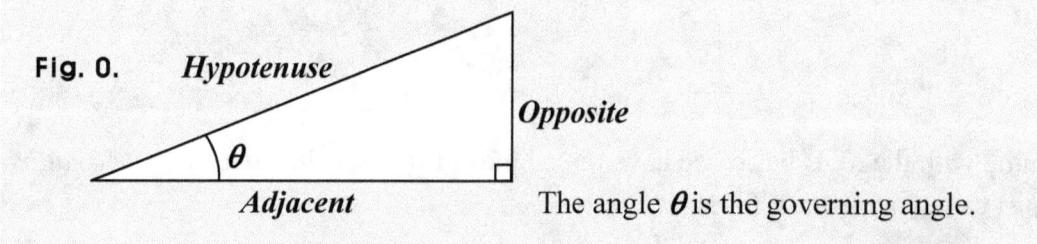

Fig. 0.   *Hypotenuse*

Opposite

θ

Adjacent   The angle *θ* is the governing angle.

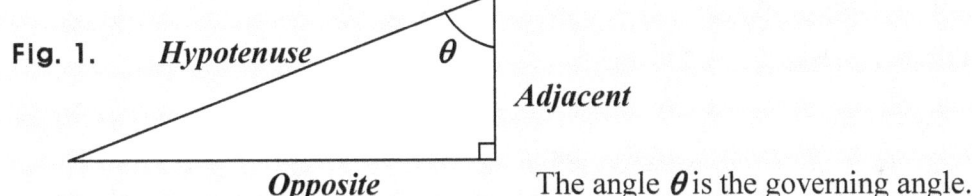

Fig. 1.   *Hypotenuse*   *θ*

Adjacent

Opposite   The angle *θ* is the governing angle.

How then, do we get each ratio?

Of the three basic trig-ratios, one is between the opposite and the hypotenuse, and is in fact, **the opposite over the hypotenuse**.

Another is between the adjacent and the hypotenuse.
That is to say that it is **the adjacent over the hypotenuse**.

And the other is between the opposite and the adjacent.
In other words, it is **the opposite over the adjacent**.

More specifically:

One ratio is called *sine*, which is denoted by **sin**, and is the ratio of the opposite to the hypotenuse, that is, **the opposite / the hypotenuse**.

Another is called *cosine*, which is denoted by **cos**, and is the ratio of the adjacent to the hypotenuse, that is, **the adjacent / the hypotenuse**.

And the other is called *tangent*, denoted by **tan**, which is the ratio of the opposite to the adjacent, that is, **the opposite / the adjacent**.

So the tangent explains how the hypotenuse is slanted, an thus, can tell us **the slope** of the hypotenuse in the right triangle.

And the tangent is in fact, no other than **the rise over the run**, that is, the slope of a line, which overlaps the hypotenuse of the right triangle. In other words, the tangent is the slope of the line that includes the hypotenuse.

What then, about the governing angle?

Suppose now, the governing angle is θ, read as theta.
Then, we put the three basic trig-ratios the way below:

- **sin θ** is the ratio of the **opposite** to the hypotenuse.

- **cos θ** is the ratio of the **adjacent** to the hypotenuse.

- **tan θ** is the ratio of the **opposite to the adjacent**.

And we read **sin θ** as <u>sine of θ</u> or just <u>sine θ</u>, for short.
We read **cos θ** as <u>cosine of θ</u> or just <u>cosine θ</u>, for short.
And we read **tan θ** as <u>tangent of θ</u> or just <u>tangent θ</u> or simply, <u>tan of θ</u>, for short.

So for instance:

We read **sin 30°** as sine of 30° or just sine 30°.
We read **cos 45°** as cosine of 45° or just cosine 45°.
And we read **tan 60°** as tangent of 60° or just tan of 60°.

And thus:
The **sine** of the governing angle is the ratio of **the opposite** to the hypotenuse.
The **cosine** of the governing angle is the ratio of **the adjacent** to the hypotenuse.
And the **tangent** of the governing angle is the ratio of **the opposite to the adjacent**.

What then, do we mean by a trig-ratio?

Assuming the governing angle is θ, we can say that:

- **sin θ** is the size of the **opposite** *relative* to the size of the hypotenuse.

- **cos θ** is the size of the **adjacent** *relative* to the size of the hypotenuse.

- **tan θ** is the size of the **opposite** *relative* to the size of the **adjacent**.

In a right triangle in fact, where the governing angle is 30°, the size of the opposite is half the size of the hypotenuse.

And taking the actual size of an object using the ratio of the object to another object, we multiply by the ratio, the actual size of the other object.

In short, finding an object using the ratio of the object to another object, we multiply the other by the ratio.

So for instance, we can get the opposite using the sine and the hypotenuse.   How?

We can get the opposite multiplying the hypotenuse by the sine.

So assuming <u>the governing angle is $\boldsymbol{\theta}$</u>, we can get the opposite multiplying the hypotenuse by **sin** $\boldsymbol{\theta}$.

And the same is true, too, for all the other trig-ratios.

So we can get the adjacent multiplying the hypotenuse by **cos** $\boldsymbol{\theta}$.

And we can get the opposite, too, multiplying the adjacent by **tan** $\boldsymbol{\theta}$, where $\boldsymbol{\theta}$ is the governing angle.

What angle then, is the governing angle?

As stated earlier, depending on the way we look at the right triangle, either of the two legs can be the adjacent. And using a wrong side as the adjacent, we get wrong ratios. So it is crucial to choose the adjacent correctly. And the adjacent is the leg *next to*, that is, *adjacent to* the governing angle.

And thus, what matters is the governing angle.

That is to say that the choice of the adjacent matters.

So assuming **P** is the adjacent in the right triangle below, we use **α** as the governing angle, and can put the three basic trig-ratios the way as follows:

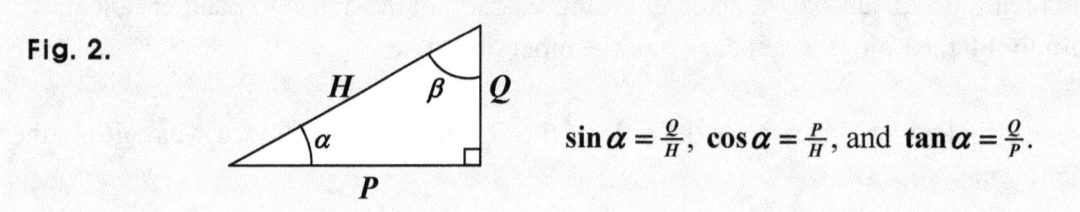

Fig. 2.

$\sin\alpha = \frac{Q}{H}$, $\cos\alpha = \frac{P}{H}$, and $\tan\alpha = \frac{Q}{P}$.

And for another instance, assuming **β** is the governing angle in the right triangle below, we have to use **q** as the adjacent, and put the three basic trig-ratios the way as follows:

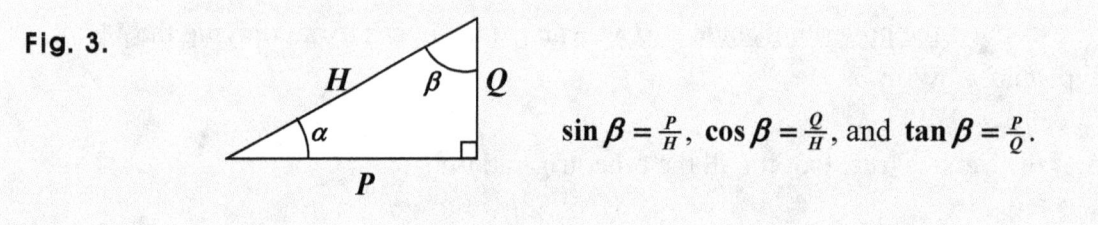

Fig. 3.

$\sin\beta = \frac{P}{H}$, $\cos\beta = \frac{Q}{H}$, and $\tan\beta = \frac{P}{Q}$.

And we have three popular governing angles, which are 30°, 45°, and 60°. Why popular?

We can easily get the trig-ratios for those angles.   How?

We can readily get them using two isosceles triangles.

One is a regular (equilateral) triangle, where every angle is 60°, and the other is a right isosceles triangle, where two of the three angles are equal, and thus, are 45° each.

So to begin with, cutting in half a regular triangle, we can get a right triangle as below:

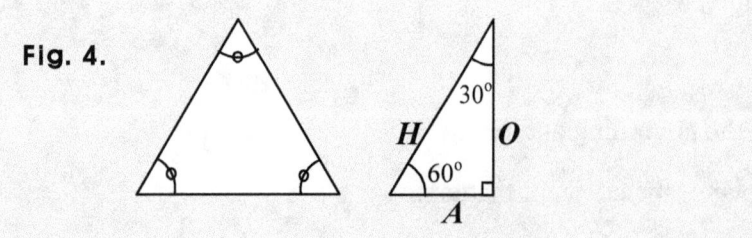

Fig. 4.

Then, noticing: $H = 2A$, and using the distance formula, we can get:

$A^2 + O^2 = H^2 = 4A^2 \Rightarrow O^2 = 3A^2 \Rightarrow O = \sqrt{3}A.$ So we get:

Fig. 5.

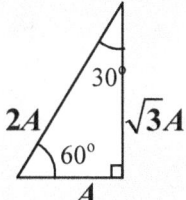

And thus, we get:

$\sin 60^o$ = the opposite / the hypotenuse $= \frac{\sqrt{3}}{2}$. $\sin 30^o$ = the opposite / the hypotenuse $= \frac{1}{2}$.

$\cos 60^o$ = the adjacent / the hypotenuse $= \frac{1}{2}$. $\cos 30^o$ = the adjacent / the hypotenuse $= \frac{\sqrt{3}}{2}$.

$\tan 60^o$ = the opposite / the adjacent $= \frac{\sqrt{3}}{1}$. $\tan 30^o$ = the opposite / the adjacent $= \frac{1}{\sqrt{3}} = \frac{\sqrt{3}}{3}$.

And next, assuming A is each of the two legs of a right triangle isosceles, we can say A is the adjacent, and can put the triangle the way below:

Fig. 4.

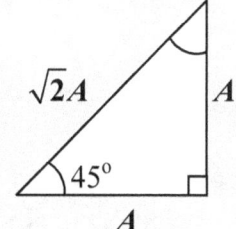

So we get:

$\sin 45^o$ = the opposite / the hypotenuse $= \frac{1}{\sqrt{2}} = \frac{\sqrt{2}}{2}$. That is, $\sin 45^o = \frac{1}{\sqrt{2}} = \frac{\sqrt{2}}{2}$.

$\cos 45^o$ = the adjacent / the hypotenuse $= \frac{1}{\sqrt{2}} = \frac{\sqrt{2}}{2}$. That is, $\cos 45^o = \sin 45^o = \frac{1}{\sqrt{2}} = \frac{\sqrt{2}}{2}$.

$\tan 45^o$ = the opposite / the adjacent $= \frac{1}{1} = 1$. That is, $\tan 45^o = 1$.

16

• What then, about the reciprocal of each?

The reciprocal of the sine is called the *cosecant*, denoted by **csc**, and is the ratio of the hypotenuse to the opposite since the sine is the ratio of the opposite to the hypotenuse.

The reciprocal of the cosine is called the *secant*, denoted by **sec**, and is the ratio of the hypotenuse to the adjacent since the cosine is the ratio of the adjacent to the hypotenuse.

And the reciprocal of the tangent is called the *cotangent*, denoted by **cot**, and is the ratio of the adjacent to the opposite, for the tangent is the ratio of the opposite to the adjacent.

And we read **csc θ** as cosecant of θ or just cosecant θ or cosec θ, for short.

We read **sec θ** as secant of θ or just secant θ or sec θ, for short.

And we read **cot θ** as cotangent of θ or just cotan of θ or cotan θ, for short.

• What then, about the inverses?

The multiplicative inverses of the basic trig-ratios are the reciprocals of the basic ones. So the multiplicative inverse of **sin θ** is **csc θ**, which is the reciprocal of **sin θ**.

• Saying however, just **the inverse of a trig-ratio**, we mean **an angle**.

Thus for instance, the inverse of **sin θ** is **θ**, and the inverse of **cos 60°** is **60°**.

And the inverse of the sine is denoted by **sin^{-1}**, which can be read as the sine inverse.

Usually though, it is read as the *arc sine*.

For instance, we know: **sin 30°** is $\frac{1}{2}$.   So **sin^{-1}$\frac{1}{2}$** is **30°**.

And **sin^{-1}$\frac{1}{2}$** is read as the arc sine of $\frac{1}{2}$.

And of course, the same is true for the other trig-ratios, too.

So the inverse of the cosine is denoted by **cos^{-1}**, read as the *arc cosine*, and the inverse of the tangent is denoted by **tan^{-1}**, read as the *arc tangent* or just the *arc tan*.

For instance, the arc cosine of $\frac{1}{2}$ is **cos^{-1}$\frac{1}{2}$**, which is 60°, and the arc tan of 1 is **tan^{-1} 1**, which is 45°. And the same is true, too, for the reciprocals: **csc**, **sec**, and **cot**.

So the *arc cosecant* is **csc^{-1}**, the *arc secant* is **sec^{-1}**, and the *arc cotangent* is **cot^{-1}**. And each of them is an angle, too, of course.

0.2. **Intro 3**

What then, can we do with trig-ratios?

Trigonometry has much to do with objects in connection with angles, that is, directions.

And we have a special object that has an amount and a direction, and call it a *vector*.

If for instance, a vector is in a 2-D space as the *x-y* plane, it can be said to have two *components*. One is called the *horizontal component*, which is parallel to the *x*-axis, and the other is called the *vertical component*, which is parallel to the *y*-axis.

- And getting each *component*, we can use a trig-ratio.

Assuming θ is the angle between the vector and the *x*-axis, we can get the vertical component using the ratio called **sin θ**, and the horizontal component using **cos θ**.

Also, we can get the slope of the vector, because the slope is the ratio, **tan θ**.

So for instance, looking at each of the right triangles presented in previous section, we can notice that the **hypotenuse** can be taken as a vector, so we can take the adjacent as the horizontal component, and take the opposite as the vertical component.

How do we get the components, though?

Fig. 0.

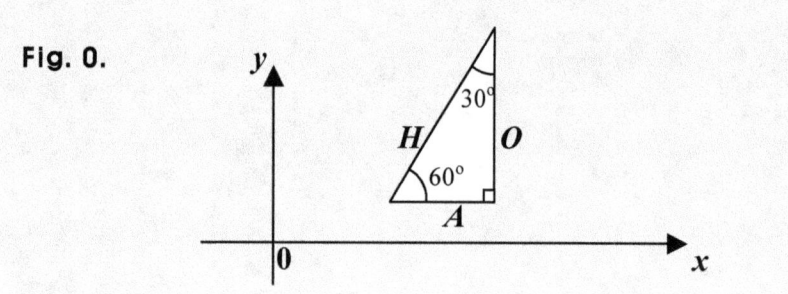

In the triangle above, *A* is the adjacent, and is the horizontal component.

And applying **cos 60°** to the hypotenuse *H*, we get: $H \cdot cos\ 60° = H \cdot \frac{A}{H} = A$.

And applying **sin 60°** to the hypotenuse *H*, we get: $H \cdot sin\ 60° = H \cdot \frac{O}{H} = O$.

So what can we do with such applications?

Suppose for instance, in the figure above, *O* is a part of a wall, a force is applied to the wall in the direction of *H*, and the length of *H* indicates the amount of the force.

Then, the amount of force acting on the wall vertically is *A*, and the amount of force acting on the wall horizontally is *O*. That is, of the force *H*, the vertical component acting on the wall is *A*, and the horizontal component acting on the wall is *O*.

So assuming for instance, *H* is 10 (kg·m/sec²) acting on the wall at 30°, and referring to the figure above, we get:

The vertical component = $H \cdot cos\ 60° = 10 \cdot \frac{1}{2} = 5$, which is *A*.

The horizontal component = $H \cdot sin\ 60° = 10 \cdot \frac{\sqrt{3}}{2} = 5\sqrt{3}$, which is *O*.

And checking to see if they are right, we get: $A^2 + O^2 = 25 + 25 \cdot 3 = 25 \cdot 4 = 10^2 = H^2$.

Why is though, the governing angle not 30° but 60°?

The governing angle is the angle between the hypotenuse and the **adjacent**, which is crucial. So getting the adjacent wrong, we get a wrong ratio. And thus, we want to make sure which side is the adjacent choosing the governing angle or applying the trig-ratios.

Examining however, the triangle above, we can notice that **sin 30°** is equal to **cos 60°**.

And in fact, if two angles are complementary to each other, the sine of one angle equals the cosine of the other. That is, if $\theta_1 + \theta_2 = 90°$, we get: **cos θ_1 = sin θ_2**. In other words, we have: **cos θ = sin (90° − θ)**, and of course, we have: **sin θ = cos (90° − θ)**, too.

And we call such an equality a trig-identity. And of course, we should be able to produce the proof, which will be therefore, coming up in one of the sections that follow.

By the way, if $\theta_1 + \theta_2 = 180°$, the two angles are said to be supplementary to each other.

• Now in large, we can say there are two categories in trigonometry: *static* and *dynamic*.

Static trigonometry begins with a right triangle, and stays with a right triangle.
So static trigonometry remains within a right triangle, which is normal or ordinary.
Why normal though? Is there then, a right triangle abnormal, too?

Like any other object in math, a triangle is an idea, too.
Expanding the idea of a right triangle, we get into trigonometry that can be said to be dynamic. In the dynamic trigonometry, we can use right triangles not only normal but *transcendental*, too.   What do we mean by though, a right triangle transcendental?

In such a case, transcendental does not mean non-algebraic, but means supernatural or beyond common thought. So it can be said to be paranormal, and thus, is certainly not a normal triangle, of course.   What then, do we mean by triangles normal?

In a normal triangle, we cannot give a non-positive value to the length of any side. If it's normal, any of the sides cannot have a length negative or 0.

What then, about triangles transcendental?

A transcendental right triangle is still a right triangle, but can have a side, the length of which is not positive. More specifically, if the side is not the hypotenuse, the side can have a length not only positive but negative or 0, too. That is, the length can have all kinds of values. So such a right triangle is beyond normal, and thus, is transcendental.

And thus, in a right triangle transcendental, the adjacent and the opposite can be positive, 0, or negative. So we can give any real number to the adjacent and the opposite.

Now, using a right triangle transcendental, we run dynamic trigonometry. So running trigonometry dynamic, we use a transcendental right triangle, where we can use any real number as the adjacent and as the opposite.   How then, do we get such a triangle?

Let's get back to the ray that keeps turning about the origin in the *x-y* plane.

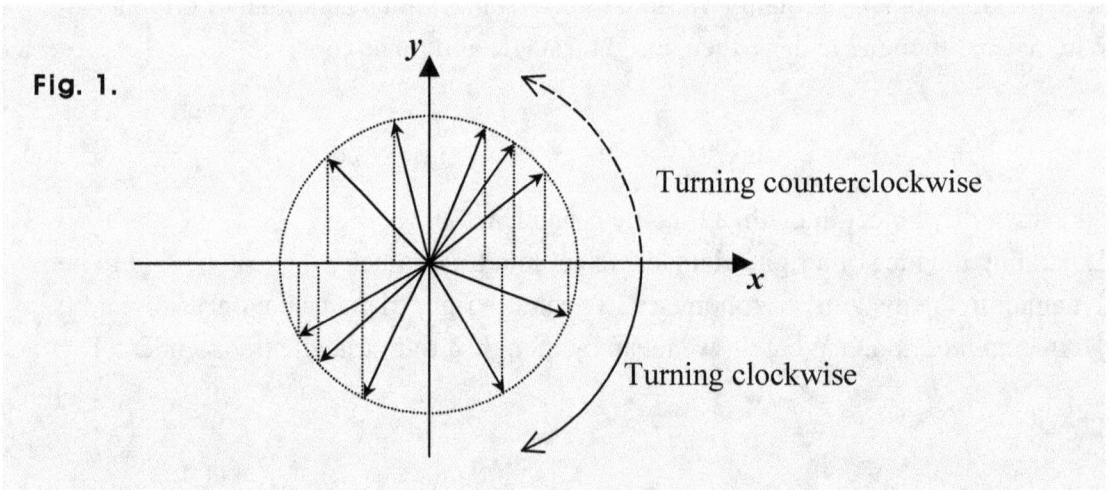

Fig. 1.

Turning counterclockwise

Turning clockwise

We can see in the figure above, quite a few right triangles. What is the adjacent in each triangle though?

In each triangle, we can use an x-coordinate as the adjacent. What then, is the opposite?

We can use a y-coordinate as the opposite. What then, is the point with the coordinates?

It is the endpoint in a line segment used as the hypotenuse, where the other endpoint is at the origin. What then, is the hypotenuse?

It is the ray that keeps turning about the origin in the x-y plane above. So the point that has the coordinates is the terminal point of the ray. And the length of the ray stays the same while the ray is turning about the origin. So the terminal point is making a circle, where the radius is the length of the ray, and of course, the center is at the origin.

And thus, assuming the ray is the hypotenuse in a right triangle, the x-coordinate at the terminal point is the adjacent, and the y-coordinate is the opposite, we can see that a right triangle keeps changing as the ray keeps turning.

However, though the right triangle changes, it remains a right triangle, that is, it just becomes a different right triangle.

So in this case, we are working with a right triangle dynamic. And thus, getting trig-ratios from such a right triangle, and applying or working with the ratios, we can say that we are running dynamic trigonometry.

Now, when the ray is on either of the coordinate axes, or is in a quadrant other than the first one, either or both of the coordinates at the terminal point cannot be positive.

That is, either of the adjacent and the opposite or both cannot be positive, which cannot happen in a normal right triangle. So the right triangle being made while the ray is turning can be said to be transcendental.

24

Like any other object in math, a triangle is an idea, too, and thus, is not a material object. Working in math, we are making and working with ideas, and what we are mainly doing is reasoning rather than calculating.

Calculations themselves are processes, and can be done by machines, too, as calculators or computers. Reasoning can however, be done by us humans only.

Working in math, we are working in idea-world as well as material world. That's because everything in math is not a material object but an idea object, which is an idea.

What then, is the purpose of the idea called a right triangle transcendental?

We know an angle is an amount of turning of the ray described above. If it turns counterclockwise, the angle is positive. And the ray can turn clockwise, too. Then, the angles made are negative. And of course, no turning means that the angle is 0°.

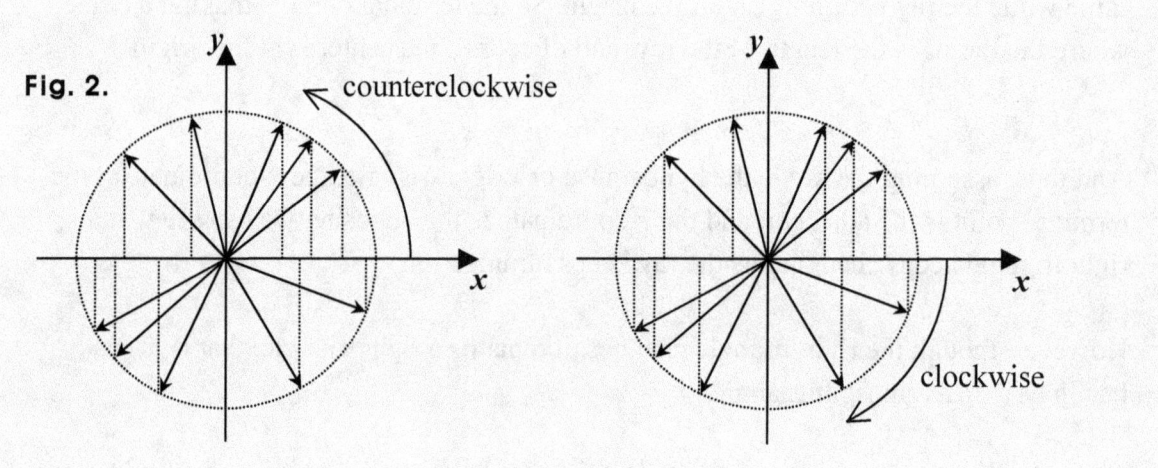

Now, we know trig-ratios have governing angles. And another major difference between the static and the dynamic is in governing angle.

In static trigonometry, we cannot use all angles as governing angles.   Why not?

A governing angle is the angle between the adjacent and the hypotenuse in a right triangle. And one angle in it is 90°, so the sum of the other two angles is 90°. One of the other two is a governing angle. And thus, a governing angle is between 0° and 90°.

In dynamic trigonometry however, we can use as a governing angle any angle made by any turning of the ray. It can turn clockwise and counterclockwise. And also, before it turns, the angle is $0°$. So a governing angle can be $0°$ or any angle positive or negative. In other words, we can use all angles as governing angles for trig-ratios.   So what?

We can put angles in a different system where we can specify angles.
The system is called *radian* system, where we use as the basis the circumference of a unit circle, which is 2π where π is the circular ratio.

(The length of an arc where the radius is r is: $r\theta$ where θ is in radian, and is the angle that the arc has, that is, the angle of the arc.)

And what's important is that using the radian system, called radian, for short, we can use all real numbers as angles that can be used as governing angles.

And in fact, what's more important is that working in trigonometry dynamic, we can come up with another set of tools, called *trigonometric functions*, often just called *trig-functions*, for short.

A trig-function is a function of an angle, which is a governing angle, and thus, takes an angle as an input. And for instance, we can use **sin** θ as a function, so assuming f is such a function, we can set: $f(\theta) = $ **sin** θ.

And of course, we can use as the variable any of other letters as x and t.

So for instance, we can set: $f(x) = $ **sin** x, where x gets an angle. What angle though?

It is a governing angle. And it is important to note that in $f(x) = $ **sin** x, the x takes a governing angle, that is, x is a governing angle, and also, that such a governing angle is the amount of angle made by the terminal ray turning about the origin in the *x-y* plane.

So what's important in the dynamic trigonometry is not only the fact that we can use all angles for trig-ratios but also the fact that the hypotenuse is constant, and another fact that the governing angle is the angle made by the ray turning.

Is the governing angle then, not the angle between the adjacent and the hypotenuse?

Suppose θ is the angle made by the ray. Then, θ is the governing angle, of course, and only if $-90° < \theta < 90°$, θ is the angle between the adjacent and the hypotenuse; otherwise, it's not. We'll get to see more about the fact when we get to the details of trig-functions.

So for now, just keep in mind that the inputs of trig-functions are governing angles, which are made by the ray turning in the *x-y* plane. What then, about outputs?

Each output is a number. What number though?

It is a trig-ratio, because *x* is a governing angle, and **sin** *x* is a trig-ratio.

That is, for each value of *x*, *f(x)* is a trig-ratio. In short, *f* gets a trig-ratio.

So in sum, all the values of *x* are governing angles, and all the values of *f* are trig-ratios.

Every value of *x* is an angle, and every value of *f* is a trig-ratio, which is a number.

Usually though, we take values of *x* as numbers, too.

That's because we often use as inputs angles expressed in radian. For instance:

$0°$ is 0 radian, which is put this way, too: 0 rad, for short.

$180°$ is π rad, where π is the circular ratio, which is 3.141592…

$360°$ is 2π rad. So what is $90°$ in radian?

It is $\pi/2$ rad. And thus, $60° = \pi/3$ rad, $-45° = -\pi/4$ rad, $30° = \pi/6$ rad, $-15° = -\pi/12$ rad, etc.

Normally however, we omit 'rad'. So we usually put angles the way below:

$90° = \pi/2, -60° = -\pi/3, 45° = \pi/4, -30° = -\pi/6, \pi = 180°, -2\pi/3 = -120°, 3\pi/2 = 270°$, etc.

What then, is $1°$ in radian?

We have: **$\pi = 180°$**, so **$1°$** is **$\pi/180$**, which is $0.01745\ldots$ What then, is 1 rad in degrees?

We have: **$\pi = 180°$**, so **1 rad** is **$180°/\pi$**, which is $57.1957795\ldots$ degrees.
And thus, **1 rad $\approx$ 57°**.

So we get: **$-360° \leq \theta \leq 360° \Leftrightarrow -2\pi \leq \theta \leq 2\pi,$ $-720° \leq \theta \leq 720° \Leftrightarrow -4\pi \leq \theta \leq 4\pi,$**   etc.

What do we mean by though, $-360°$ or $-720°$?

If the ray makes a complete turn clockwise, it makes $-360°$, and if it makes one and a half complete turns clockwise, it makes an angle of $-360°\cdot1.5$, which is $-540°$.

And we know $\pi, \pi/2, \pi/3, \pi/4, \pi/6, 2\pi/3, 3\pi/2$, etc. are all numbers.   So what can θ be?

The ray can make as many turns as necessary either counterclockwise or clockwise.

So we can put it this way: **$-360°\cdot n \leq \theta \leq 360°\cdot n \Leftrightarrow -2n\pi \leq \theta \leq 2n\pi$**, where **$n$** is an integer.

That is, we can get: **$-\infty < \theta < \infty$**, where ∞ indicates infinity. So θ can be all real numbers.

And thus, we usually take values of x as numbers, too, which are in nature however, angles. So keep in mind that input values of a trig-function are actually angles.

Now, in sum, we have two categories in trigonometry.

One is static trigonometry. In trigonometry static, we use normal right triangles applying or working with trig-ratios. A normal right triangle has a side called the hypotenuse, together with two sides called the adjacent and the opposite perpendicular to each other. And of course, the lengths of all the sides are positive.

And in dynamic trigonometry, we mainly work with trig-functions. In trigonometry dynamic, we use a transcendental right triangle, which keeps changing, and thus, is dynamic. A right triangle transcendental, too, has a side called the hypotenuse, together with two sides called the adjacent and the opposite perpendicular to each other. However, either or both of the adjacent and the opposite can be negative or 0 as well as positive.

And in trig-functions, we have three basic ones as follows:

One is a sine function, which is: **sin** x. Another is a cosine function, which is: **cos** x. And the other is a tangent function, which is: **tan** x, where x usually takes angles in radian.

And usually, we just take as numbers *angles in radian*.

So for instance, assuming the domain is a set of all angles, and f is a sine function, we can put the trig-function f the way as follows: $y = f(x) = $ **sin** x for x real.

What do we mean by 'x real' though?

It means that the input variable can take all real numbers, that is, any real number. So the domain is a set of all real numbers.

And putting angles in radian, we express angles by means of numbers.

For instance, $\pm 30° = \pm \pi/6$, $\pm 45° = \pm \pi/4$, $\pm 60° = \pm \pi/3$, $\pm 90° = \pm \pi/2$, and $\pm 180° = \pm 2\pi$.

And in the function f above, x takes in fact, an angle. So 'x real' means that x can take any angle. And thus, the domain is a set of all angles.

Normally though, we say that the domain is a set of all real numbers if we use radians as angles.

What then, about the range?

The range is a set of all numbers from -1 to 1, and each of the numbers is a trig-ratio.

So the range of f can be put this way, too: **-1 ≤ y ≤ 1**, or **|y| ≤ 1**.

And the sine function f above can be called the prototype, which is thus, in the most basic form. And assuming g is a sine function, too, where the domain is a set of all angles, and using a general form, we can put it the way below:

$y = g(x) = A \cdot \sin\{w(x + a)\} + b$ for x real, where A, w, a, and b are constant.

And for short, we can just put it this way, too: $y = g(x) = A \cdot \sin w(x + a) + b$ for x real.

Then, the range of g is a set of all numbers from $-|A| + b$ to $|A| + b$.

So the range of g can be put this way, too: $-|A| + b \leq y \leq |A| + b$, or $|y| \leq |A|$.

What do we mean by though, $w(x + a)$?

Setting: **sin θ = sin $w(x + a)$**, we get: **θ = $w(x + a)$**, so $w(x + a)$ represents an angle.

What then, about A and b?

They represent a number each. So each of the two takes a number.

What then, about the range of this function: $y = h(x) = A \cdot \sin w(x + a)$ for x real?

The range of h is a set of all numbers from $-|A|$ to $|A|$.

So the range of h can be put this way, too: $-|A| \leq y \leq |A|$, or $|y| \leq |A|$.

And in both cases of g and h, $|A|$ is called the *amplitude*.

And $|w|$ is called the *frequency*, $\frac{2\pi}{|w|}$ is the *period*, and a is called the *phase*. And we will get to the details on A, w, a, and b in one of the sections where we discuss trig-functions.

Now, we have been though the overview of trigonometry.

So let's next, get into the details on each of the objects and tools been introduced. That is to say that, in the coming sections, in each section, you are going to see the detailed explanations on each and every concept introduced in the previous two sections and this section, you will get to understand all the expressions and components of the trig-functions introduced in this section.

If you want to get the clear picture of all the ideas and concepts though, you need to first, get a clear picture of angles, that is, you need to understand the concept of an angle first. An angle is a math object, and is not just some degrees in math. It is from another math object called a circle. If we didn't have a circle, then we had no angle.

So the next section begins with the ideas of circles and angles.

1. Circles and Angles

So to begin with, what is a circle?

A circle is a 2-D object, which is a collection of all points that are the same distance away from a particular point. And the particular point is called the center of the circle, and the same distance is called the radius of the circle. Also, a circle is said to have its diameter, which is a line segment passing through the center and connecting two points facing each other in the circle. So the diameter is twice the radius of the circle.

What then, about the circumference of a circle?

It is the length of the curved and closed line segment forming the circle. For instance, assuming the radius is *r*, and *C* is the circumference, we get: $C = 2\pi r$, where π is the circular ratio, which is an irrational number, and is 3.141592…

Why is though, the same distance stated above called the radius?

Saying *the* radius of a circle, we mean the distance from the center to a point in the circle.

That's because taking any of such distances in a circle, we get one distance only.

Saying however, just a radius of a circle or a radius in a circle, we mean a line segment connecting the center and a point in the circle.   How many points are there in a circle?

Infinitely many, of course. So the center of a circle can be said to *radiate* infinitely many line segments, each of which has the same length. And thus, probably for that reason, we call each of the line segments a *radius*, and call the same length *the* radius.

Suppose now, all the radii in a circle are terminal rays, that is, rays with finite lengths, and all the rays begin at the center of the circle.

Then, all their lengths are equal, so all their terminal points are in the circle. That is to say that all the terminal points form the circle. And we can say that the center emits such rays in all directions.

Fig. 0

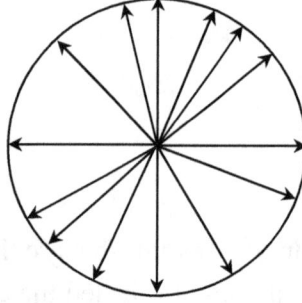

Also, we can say that each of all the rays has a different direction. What then, do we call the difference between the directions of two rays?

We call it an angle. More specifically, we call it the angle between the two rays. And thus, we can say that *a difference in direction is an angle*. And we know that the rays are the radii in the circle. So we can say that every radius in a circle has a different direction. What then, can we say a pair of radii can form?

Two radii in a circle can be said to form a geometric object, which can show a difference in direction.   What geometric object though?

Two radii in a circle have different directions. And a difference in direction is an angle.

Fig. 1

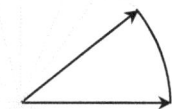

So as shown above, an angle can be indicated by a circular wedge made of two radii and a part of a circle. And thus, a pair of radii in a circle can be said to form a geometric object called an angle.   Is there anything else though, we can call an angle?

Suppose next, a terminal ray is turning about the origin in the *x-y* plane.
What then, do we call the amount of turning the ray makes?

We call it an angle, too. So we can say that an angle is an amount of turning, too.

Fig. 2

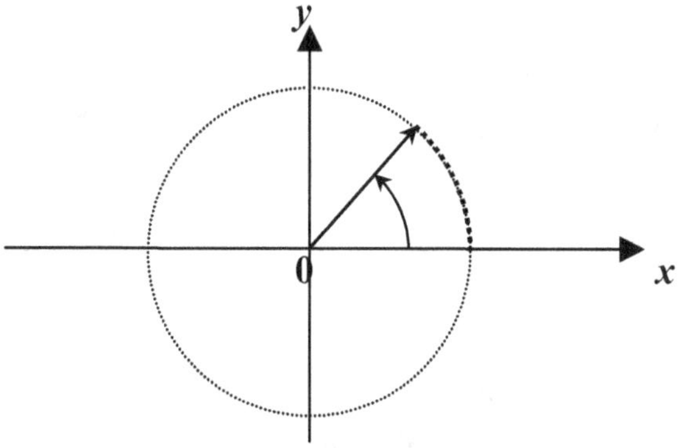

And also, we can say a difference in direction is an amount of change in direction.
So in short, an angle can be called a change in direction, too.

And thus, we can use an angle specifying an amount of turning or a change in direction. In short, *an angle* is *an amount of turning* or *a change in direction.*

(And we know a pair of radii in a circle form an angle, which is a geometric object. So we can put an amount of angle this way, too: An amount of angle can tell us how much two radii in a circle are apart from each other. For instance, if in a circle, a radius is 30° away from another radius in the circle, the angle between the two radii is 30°.)

And of course, expressing an amount of angle, we use a unit of measure as in the cases of other objects as lengths, areas, volumes, weights, etc. For instance, we use kilograms or pounds for weights, liters or gallons for volumes, meters or miles for lengths, and square meters or acres for areas, etc.   What then, is the unit of measure for angles?

We have two kinds in such a unit. One is *degree*, and the other is *radian*, called *rad* for short. And thus, we can use degrees or radians for angles. So for instance, an angle can be 25 degrees, 1 degree, 3 radians or 3 rad, or 1 radian or 1 rad.

And using degrees and putting in writing an amount of angle, we use a symbol, which is a small circle, and we put it at the upper right-hand corner of the number indicating the amount of angle, that is, we use it as a superscript as in 25°.   What then, about radians?

We use no symbol for radians. So assuming for instance, A is known to be an angle in radian, we just put it in writing this way: $A = 2$, which means A is 2 radians.

Having to clarify though, an angle is in radian, we put 'rad' after the number. So for instance, if an angle B is 3 radians, we can put it in writing this way: $B = 3$ **rad**.

And thus, we have two metrologies for angles, that is, two systems of measurement for angles. One is radian system, and the other is degree system. And one system can get converted to the other. So two different amounts in angle can mean the same angle. For instance, we have: $180^\circ = \pi$ **rad** where π is the circular ratio, which is 3.141592… And briefly, we put it this say: $180^\circ = \pi$.   How can we calculate though, an amount of angle?

As explained above, an angle is a geometric object, which is a wedge formed by two radii in a circle and a part of a circle.   What then, do we call such a part?

We call it an arc, which is a figure that can show not only a difference (or a change) in direction but an amount of turning, too.

Fig. 3
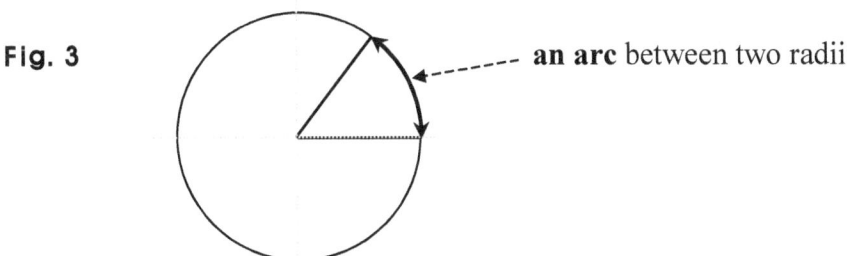
an arc between two radii

And we can calculate an amount of angle by means of an arc and a circle.
More specifically, we can get an amount of angle using a ratio of an arc length to the circumference of the circle the arc belongs to.

• In short, an angle can be put in terms of a ratio of an arc to the circle that has the arc.

How come though?

Let's get back to the ray turning about the origin in the *x-y* plane.

Fig. 4
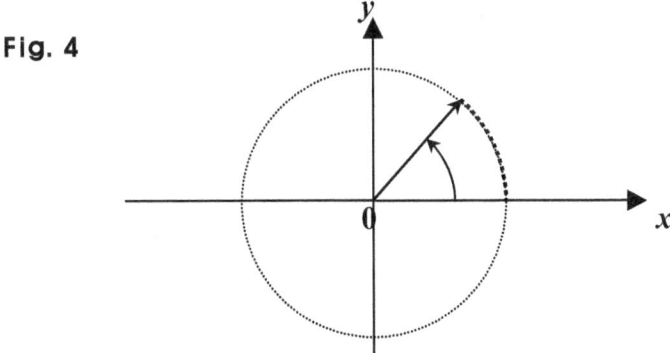

Then, first, an angle is an amount of turning. And an arc is said to have an angle.

An amount of angle is however, not the length itself of an arc, because there can be many, or rather, infinitely many arcs between two radii in a circle. In the figure above, one of the two is the ray turning, and the other is on the *x*-axis.

And next, as the ray turns, the arrowhead makes an arc, and if the ray makes a complete turn, the arrowhead makes a circle. So if the ray makes a complete turn, the length of the arc made is the circumference of the circle, that is, the arc is the circle itself. And if a half of a complete turn is made, the arc is half the circle. If a third of a complete turn is made, the arc is a third of the circle.

So an amount of turning, that is, the angle that an arc has is proportional to the ratio of the length of the arc to the circumference of the circle the arc belongs to. That is to say that the angle an arc has is proportional to the ratio of the arc to the circle that has the arc. So in short, an angle is proportional to the ratio of an arc to a circle.

And thus, assuming θ is an angle, k is such a ratio, and C is a constant, we can set: $\theta = Ck$. What then, is the constant C?

It is 360°.   How come though?

If the ray makes a complete turn, that is, the ray gets for the first time, the same direction as the direction it had before turning, in other words, if the arc made is a circle, we say that the angle made is 360°. So in short, a complete turn is 360°. And thus, if the ray has made two complete turns, that is, if the ray gets for the second time, the same direction as the direction it had before turning, the angle made is twice 360°, that is, 720°.
What then, about a half of a complete turn?

If the ray gets for the first time, the direction opposite of the direction it had before turning, the ray makes a half of a complete turn, that is, the arc made is a half circle, so the angle made is a half of 360°, that is, 180°. So in short, a half turn is 180°. And in turn, a quarter turn is: $360^\circ/4 = 90^\circ$.   What then, about one and a half turn?

If the ray gets for the second time, the direction opposite of the direction it had before turning, it has to make another complete turn after the first half of a complete turn, so the angle made is: $180^\circ + 360^\circ = 0.5 \cdot 360^\circ + 1 \cdot 360^\circ = (0.5 + 1)360^\circ = 1.5 \cdot 360^\circ$.

Thus in short, one and a half turn is: $1.5 \cdot 360^\circ = 1.5 \cdot 2 \cdot 180^\circ$, which is: $3 \cdot 180^\circ = 540^\circ$.

What then, about a third of a complete turn?

The angle made is a third of $360°$, that is, $120°$. And thus, we can specify an angle by the product of $360°$ and the ratio of the arc to the circle the arc belongs to.

So for instance, assuming the length of an arc is 2, and 12 is the circumference of the circle where the arc belongs, we get: 2/12 of $360°$, which is 1/6 of $360°$, which is $60°$. So assuming θ is an angle, and k is such a ratio, we can set: $\boldsymbol{\theta = 360°k}$.

- And we can put an angle the way below, too:

An angle can be said to indicate how much two lines meeting at a point are apart from each other. If two lines are parallel to each other, the angle is 0. What if two line segments not parallel are away from each other, and thus, do not meet at a point?

A line segment is in a line, and lines not parallel meet at a point. So the angle between the two lines the two line segments belong to is the angle between the two line segments.

Fig. 5

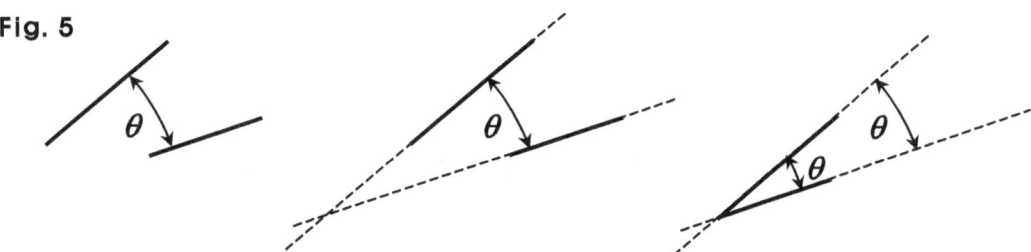

What then, about $1°$?

Dividing a circle into 360 equal parts around the center, we get 360 equal wedges. So taking one of the wedges, we get: $360°/360 = 1°$. So if the ray tuning counterclockwise makes a three hundred and sixtieth of a complete turn, we say that the angle made is $1°$.

And again, dividing one wedge worth $1°$ into 60 equal parts, we get 60 equal smaller wedges, each of which has an angle called 1 *minute* denoted by **1'**. So we get: $1°/60 = \boldsymbol{1'}$. And thus, if the ray tuning counterclockwise makes a twenty thousand and six hundredth of a complete turn, we say that the angle made is 1'.
Then again, taking 1' apart into 60 equal parts, we get an angle called 1 *second* denoted by **1"**.   So we get: $1°/360 = 1'/60 = \boldsymbol{1"}$.

And we can keep going taking the smaller wedges. We do not normally though, get a wedge worth less than 1 second, 1". What then, about 0.78°?

It is an angle that is 78% of 1°, so $0.78° = 78 \cdot 1°/100$. And thus, we get: $0.1° = 1°/10$. So the bigger the wedge is, is the bigger the angle, in other words, the bigger the arc is, is the bigger the angle?

Not necessarily. Many different arcs can have the same angle.

Fig. 6

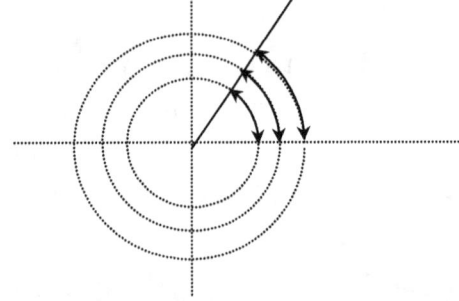

And thus, what matters in an angle is not an arc length itself but the ratio of an arc length to the circumference of the circle the arc belongs to.
That's because every circle has 360°, but the radius of each can be different. So what?

First, many arcs with different lengths can have the same angle. All those arcs belong to circles said to be concentric. Concentric circles have the same center as shown above.

Now, holding an angle constant, we can see the larger the radius, the larger the arc.
That is, for a particular angle, as the radius grows, the arc grows, too.
How then, does the arc grows as the radius grows?

Getting back to the figure above, we can see three wedges, each of which is made of two radii and an arc. And extracting them, we get:

Fig. 7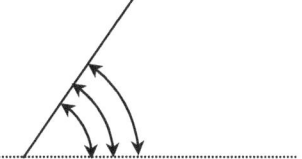

Then, the three wedges are said to be similar to each other.

So the ratio between the radii is the same as the ratio between the arcs.

For instance, assuming R_1, R_2, and R_3 are the radii, and A_1, A_2, and A_3 are the arcs, and also, assuming $R_1 < R_2 < R_3$, and $A_1 < A_2 < A_3$, we get: $R_1 : R_2 : R_3 = A_1 : A_2 : A_3$.

And thus, we can get: $A_1/R_1 = A_2/R_2 = A_3/R_3$.   What then, is the value of A_1/R_1?
In other words, if $A_1/R_1 = A_2/R_2 = A_3/R_3 = D$, what is D?

We know D is the ratio of each arc to its corresponding radius, and is constant.
And also, the angle in each of the three wedges, that is, the angle of each arc is the same, and thus, is constant. So we can reasonably expect that the ratio is the angle.
And the ratio is in fact, the angle. And thus, D is the angle.
We know however, a ratio is a number.   How come an angle can be a number?

As mentioned earlier, we have two metrologies for angles, that is, two systems of measurement for angles. One is radian system, and the other is degree system.

And using the radian system, we can use as angles all real numbers, which are therefore, angles in radian. That was covered in the previous section. And we will see shortly how we get angles in radian.

Now, in such a circular wedge, the ratio of the arc to the radius is the angle of the arc.

So assuming r is the radius, A is the arc, and θ is its angle in radian, we get: $A/r = \theta$.
That is, we get: $A = r\theta$.

And next, getting back to the concentric circles above, and holding the radius constant, that is, looking at one circle only, we can see the larger the arc, the larger its angle.

More specifically, the arc is proportional to the angle.

And thus, assuming r is the radius, A is an arc, and θ is its angle, we get: $A = r\theta$.
So for instance, if the angle is $360°$, we get: $\theta = 2\pi$, and thus, we get: $A = 2\pi r$.

Now, stating angles in terms of degrees, we put angles in degree system.
We have two systems where we can put angles. What then, is the other system?

It is called radian system, where we put angles in terms of radians. So 1 radian is a unit of measure for angles. What is 1 radian though?

Going back to one of the wedges stated above, we can get:

Fig. 8

Now, if the arc length is the radius of the circle that has the arc, the angle the arc has is 1 radian. That is, if an arc length is the radius, the angle of the arc is 1 radian.

- So in short, if an arc is the radius, the angle is 1 rad.

And thus, if a circular wedge is like a regular (equilateral) triangle where three sides are equal, the angle of the arc is 1 rad. And we can take any angle based on the idea where we get 1 radian the way stated above: if an arc is the radius, we get 1 rad. So it is probably the case where getting angles the way above, we call the angles radians.

That is, 1 radian is **radi**us **an**gle, the angle of the arc the length of which is the radius. What exactly is the way though, we get such angles in radian?

Such an angle is a ratio.

In radian system, an angle is the ratio of the arc to the radius in a circular wedge.

So we have: $A/r = \theta$ where A is the arc, r is the radius, and θ is the angle A has.

And thus, assuming $A = r$, we get: $\theta = A/r = r/r = 1$, which is called 1 radian. What angle in the degree system then, is 1 radian? That is, what degree is 1 rad?

Assuming first, the arc is a sixth of a circle of radius r, we get: $A = 2\pi r/6 = \pi r/3$, so we get: $\theta = A/r = (\pi r/3)/r = \pi/3$, called $\pi/3$ radian. And we know the angle the arc has is $60°$. So we get: $60° = \pi/3$ rad.   What then, about a quarter circle?

The length of a quarter circle of radius r is $2\pi r/4$, and the angle a quarter circle has is $90°$. So we get: $\theta = A/r = (2\pi r/4)/r = (\pi r/2)/r = \pi/2$, and thus, we get: $90° = \pi/2$.

And by the same token, since the angle of an eighth of a circle is $45°$, if the arc is an eighth of a circle, its angle θ is: $A/r = (2\pi r/8)/r = \pi/4$, and thus, we get: $45° = \pi/4$.

And also, the arc can be a half circle. Then, the circular wedge is a half of a circular disk.

So assuming $A = 2\pi r/2 = \pi r$, we get: $\theta = A/r = \pi r/r = \pi$, called π radian. And we know the length of a half circle of radius r is πr, and the angle a half circle has is $180°$. So we get: $180° = \pi$. And of course, the arc itself can be a circle, too, that is, the wedge can be a circular disk. Then, we get: $A = 2\pi r$, so its angle θ is: $A/r = 2\pi r/r = 2\pi$, and thus, we get: $360° = 2\pi$ rad, since the angle a circle has is $360°$. So next, what angle in degree is 1 rad?

We have: $60° = \pi/3$ rad, $90° = \pi/2$ rad, $45° = \pi/4$ rad, $180° = \pi$ rad, $360° = 2\pi$, etc. And using any of the ones above, we can get the angle in degree equivalent to 1 rad. Using the fourth one above, we get; $180° = \pi$ rad $\Rightarrow$ 1 rad $= 180°/\pi = 57.29657...°$. So we get: 1 rad $\approx 57°$.   What angle then, in radian is $1°$?

We have: $45° = \pi/4$ rad.   So we get: $1° = \{(\pi/4)/45\}$ rad $= (\pi/180)$ rad $= 0.01745\dots$ rad. So we get: $1° \approx 0.0175$ rad.   Why angles in degrees or radians though?

We can use either system working with angles. Working with angles though, we can say in large that we use angles in two different cases. In one case, we use angles that do not change, and in the other, we use angles that can change.

And angles that do not change can be said to be static, and angles that can change can be said to be dynamic. And using angles static, we are likely to use the degree system, and using angles dynamic, we are likely to use the radian system. In either case though, we can still use either system.

Using angles in degree, we have to use a symbol, which is a small circle, and need to put it upper right corner of the number indicating the amount of the angle as in $90°$.

Using angles in radian however, we just use real numbers as the angles. Usually, using angles in radian, we just use numbers only, and do not put rad or radian to the numbers.

So for instance, assuming A is π radian, we just put it this way: $A = \pi$, which represents a number $3.141592\dots$, which is an irrational number, which is a real number.

And for another instance, assuming B is 7 rad, we simply put it this way: $B = 7$. What then, can be the advantage of using angles in radian?

Suppose we work with a function, where the inputs are angles. Then, the input variable is an angle that can change, and thus, is dynamic (characterized by continuous change).

And it's convenient to use real numbers as angles. So in such cases, we normally use angles in radian rather than those in degree.

• Let's now get back again to the ray turning about the origin in the *x-y* plane, and look at angles from a bit different perspective.

Fig. 9

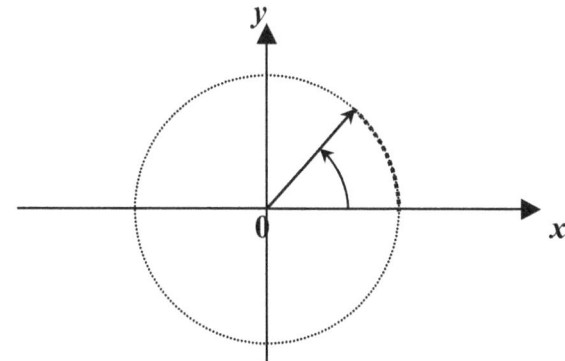

Then, we can see that the ray keeps changing its direction, and that the angle between the ray and the *x*-axis changes as the ray turns. We know that the *x*-axis is fixed. So what makes an angle is the ray turning, and thus, we can say for simplicity that the angle between the ray and the *x*-axis is the angle of the ray, which is the angle the ray makes.

And initially, that is, when the ray is resting on the *x*-axis, its angle is assumed to be 0°.

Suppose now, the ray starts turning counterclockwise.

Then, as the ray keeps turning, it keeps changing its direction, and its angle keeps changing. And in this case (tuning counterclockwise), the angle is said to be positive and increasing. And knowing the angle at a particular moment, we can specify the direction of the ray, at the particular moment, of course.   How then, do we get the angle?

Technically, a circle is an arc, too. And by definition, a circle has 360°, which is 2π rad, and a half circle has 180°, which is π in radian. So if the ray tuning counterclockwise makes a complete turn (or rotation) about the origin, its arrowhead makes a circle, and thus, we say the angle made is 360°, that is, 2π. And if two complete turns are made, the arrowhead makes a circle twice, so the angle made is twice 360°, and thus, is 720°, that is, 4π. Why not though, 360° but 720°?

The angle made in this case is the amount of turning, which is in this case, twice a complete turn. So the angle can be more than 2π, and for instance, can be 3π, $7\pi/2$, etc.

- What if the ray is turning clockwise?

Then, its angle is said to be negative. So for instance, if the ray makes a complete turn clockwise, its angle is -360°, that is, -2π rad, and if a quarter of a complete turn is made clockwise, the angle made is -90° or -π/2.

And thus, tuning the ray counterclockwise or clockwise, we can get all angles. And indicating all such angles, we can use all real numbers.   So what?

For instance, we can come up with functions where inputs are angles. And it's convenient to use numbers as inputs. So in such a function, using as inputs angles in radian, we can use all real numbers as such inputs.   What are angles for though?

Working with objects that have to do with directions or that can change their directions, we need to work with objects called angles. We don't just work with angles though, of course. The angle we get to work with in such a case is one of the angles in a triangle. It's not just any triangle though.   What triangle then?

Of all kinds in triangles, the simplest and the most basic, and thus, the most fundamental is a triangle where one angle is 90°, called a right angle. So such a triangle is called a right triangle. And a right triangle is the place where a special geometry called trigonometry begins. So in the next section, we are going to begin with ideas called triangles prior to the ideas in trigonometry.

In short:

To begin with, by definition, putting angles in degrees, we get 360° if the ray makes a complete turn.

If the ray does not make any tuning, that is, it is just sitting on the *x*-axis, the angle made is 0°. If the ray is turning counterclockwise, the angle made is said to be positive, and if the ray is turning clockwise, the angle made is said to be negative. So for instance:

If the ray is turning clockwise, and the ray makes a twelfth of a complete turn, the angle θ made is: **$-360°/12 = -30°$**.

And next, angles in radians are ratios.   What ratios?

An angle in radian is a ratio of an arc to the radius of a circle that has the arc. So assuming A is the arc, r is the radius, and θ is the angle A has, we get: $\theta = A/r$.

And such a ratio can be negative or 0 as well as positive.

If the ray does not make any tuning, that is, it is just sitting on the x-axis, the angle made is 0. If the ray is turning counterclockwise, the angle made is said to be positive, and if the ray is turning clockwise, the angle made is said to be negative, and thus, we get:

$A = r|\theta|$, because A is a length, which is the length of the arc made, of course.

Thus for instance:

If r is the length of the ray turning clockwise, and the ray makes a twelfth of a complete turn, the arc A made is: **$2\pi r/12 = \pi r/6$**, so the angle θ made is $-A/r = -(\pi r/6)/r = -\pi/6$. And we know putting the angle in degree, we get: **$-30°$**. So we get: **$-30° = -\pi/6$**. And thus, we get: **$180° = \pi$** radian.

And the arc A can be half the circle or bigger. So:

If r is the length of the ray turning, and the ray makes a complete turn either clockwise or counterclockwise, the arc A made is: $r|\theta| = 2\pi r$, since the angle θ made is **-2π** or **2π**.

And if r is the length of the ray turning, and the ray makes one and a half of a complete turn either direction, the arc A made is: $r|\theta| = 3\pi r$, since the angle θ made is **-3π** or **3π**.

Examples in Circles and Angles

0. Put the angles below in radian.

0. 30°	1. 45°	2. 60°	3. 90°	4. 120°
5. 180°	6. 270°	7. 360°	8. 405°	9. 775°

(Note that the examples below are for those students who want to take courses in calculus in university level.)

Suppose two terminal rays are turning in the *x-y* plane, and turn about the origin, of course, and both turn in the same direction. So for instance, turning counterclockwise, they keep turning counterclockwise only. Then:

1. Suppose one ray is in the second quadrant, and the angle made is θ. What quadrant then, the other ray has to be in if the angle made is $\theta/2$?

2. Suppose again, the angle made by one ray is θ, but this time, the angle made by the other ray is 6θ. Then:

Suppose they are in a straight line, and indicate opposite directions, that is, the angle between the two is π. Then:
2.0. Find the angle θ.   2.1. Find the angle θ assuming $0 < \theta < \pi/2$.

Suppose this time, they are in a straight line, and indicate the same direction, that is, the angle between the two is 0. Then:
2.2. Find the angle θ.   2.3. Find the angle θ assuming $0 < \theta < \pi/2$.

Suppose this time, they are symmetric about the *y*-axis. Then:
2.4. Find the angle θ.   2.5. Find the angle θ assuming $0 < \theta < \pi/2$.

Suppose this time, they are symmetric about the *x*-axis. Then:
2.6. Find the angle θ.   2.7. Find the angle θ assuming $0 < \theta < \pi/2$.

Suggestions or Solutions
To the **Problems** in the Example **0**

To begin with, by definition, we have: $180° = \pi$ rad.

And normally, rad is omitted, so in short, we just set: $180° = \pi$.

And we know 180 degrees is 180 of 1 degrees, that is, $180°$ is 180 of $1°$s.

In other words, we have: $180 \cdot 1° = \pi$.

So next, putting $1°$ in radian, we get: $1° = \pi/180$.

And next, we can put an angle $A°$ this way: $A \cdot 1°$.

And we know: $1° = \pi/180$.

So we can get: $A° = A \cdot 1° = A(\pi/180) = (A/180)\pi$, which is in radian.

That is to say that we get: $A° = (A/180)\pi$.

And thus, we get:

0. $30° = (30/180)\pi = (1/6)\pi = \pi/6$.

1. $45° = (45/180)\pi = (1/4)\pi = \pi/4$.

2. $60° = (60/180)\pi = (1/3)\pi = \pi/3$.

3. $90° = (90/180)\pi = (1/2)\pi = \pi/2$.

4. $120° = 2 \cdot 60° = 2(\pi/3) = 2\pi/3$.

5. $180° = (180/180)\pi = \pi$.

6. $270° = 3 \cdot 90° = 3 \cdot \pi/2 = 3\pi/2$.

7. $360° = (360/180)\pi = 2\pi$.

8. $405° = 360° + 45° = 2\pi + \pi/4 = 9\pi/4$.

9. $775° = (775/180)\pi = (155/36)\pi = 155\pi/36$, which equals $4\pi + 11\pi/36$.

Suggestions or Solutions
To the **Problem** in the Example 1

Two terminal rays are turning about the origin in the *x-y* plane. And both turn in the same direction. One ray is now, in the second quadrant, and the angle made is θ. What quadrant then, the other ray has to be in if the angle made is $\theta/2$?

Assuming first, $\pi/2 < a < \pi$, and setting: $\theta = 2n\pi + a$ where n is an integer ≥ 0, we get:

$\theta/2 = (2n\pi + a)/2 = n\pi + a/2$, and $\pi/4 < a/2 < \pi/2$.

And thus, if n odd, it is in the third quadrant.

If however, n is even, it is in the first quadrant.

Assuming next, $-3\pi/2 < a < -\pi$, and setting: $\theta = 2n\pi + a$ where n is an integer ≤ 0, we get: $\theta/2 = (2n\pi + a)/2 = n\pi + a/2$, and $-3\pi/4 < a/2 < -\pi/2$.

And thus, if n is odd, it is in the first quadrant.

If however, n is even, it is in the third quadrant.

If not quite sure of the idea behind the processes above, follow the steps below:

To begin with, can we just set: $\pi/2 < \theta < \pi$, since the ray is in the second quadrant?

We don't know how many complete turns the ray in the second quadrant has made before it made the angle θ. And also, we don't know in what direction it turns, that is, we don't know if it turns counterclockwise or not. How then, can we set the angle θ?

We can have two cases where the rays turn counterclockwise, and they turn clockwise. So suppose first, the rays turn counterclockwise.
Then first, making a complete turn, the angle made by the ray is 2π.

And next, if the ray is in the second quadrant, the angle between the ray and the x-axis on the right of the origin is between $\pi/2$ and π.

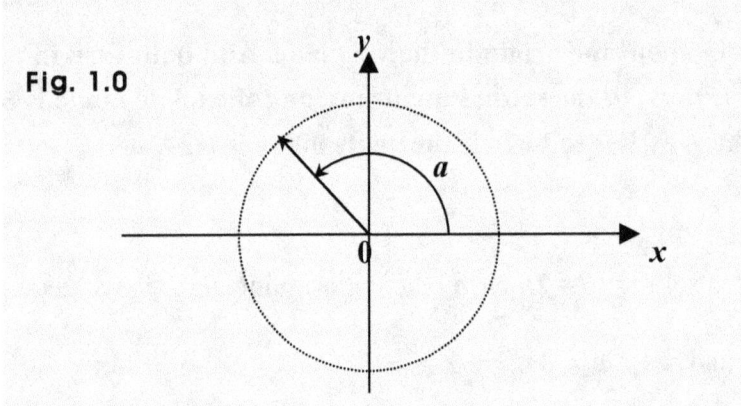

Fig. 1.0

So assuming $\pi/2 < a < \pi$, we want to set: $\theta = 2n\pi + a$ where n is an integer ≥ 0.

How then, can we set the other angle made by the other ray?

The angle made is $\theta/2$, and we have: $\theta = 2n\pi + a$. So $\theta/2 = (2n\pi + a)/2 = n\pi + a/2$.

And we know: $\pi/2 < a < \pi$.   So we get: $\pi/4 < a/2 < \pi/2$.

That is to say that $a/2$ is an acute angle.

In what quadrant then, is the other ray if n is odd?

It is in the third quadrant. And if n is even, it is in the first quadrant.

Suppose next, the rays turn clockwise.

Then first, making a complete turn, the angle made by the ray is -2π.

And next, if the ray is in the second quadrant, the angle between the ray and the x-axis on the right of the origin is between $-3\pi/2$ and $-\pi$.

Fig. 1.1

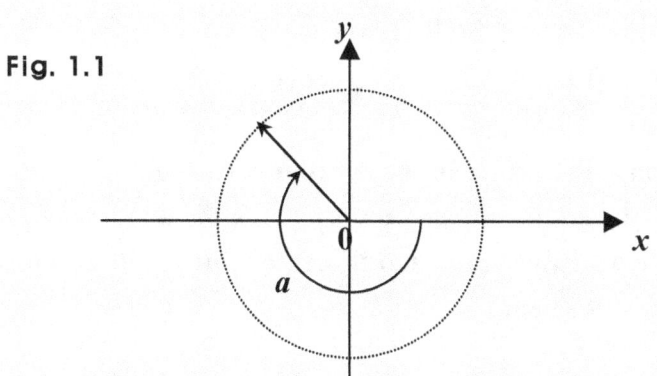

So assuming $-3\pi/2 < a < -\pi$, we want to set: $\theta = 2n\pi + a$ where n is an integer ≤ 0.

How then, can we set the other angle made by the other ray?

The angle made is $\theta/2$, and we have: $\theta = 2n\pi + a$.   So $\theta/2 = (2n\pi + a)/2 = n\pi + a/2$.

And we know: $-3\pi/2 < a < -\pi$. So we get: $-3\pi/4 < a/2 < -\pi/2$.

That is to say that $a/2$ is an obtuse angle negative.

In what quadrant then, is the other ray if n is odd?

It is in the first quadrant. And if n is even, it is in the third quadrant.

In short:

Assuming first, $\pi/2 < a < \pi$, and setting: $\theta = 2n\pi + a$ where n is an integer ≥ 0, we get:
$\theta/2 = (2n\pi + a)/2 = n\pi + a/2$, and $\pi/4 < a/2 < \pi/2$.

And thus, if n odd, it is in the third quadrant.
If however, n is even, it is in the first quadrant.

Assuming next, $-3\pi/2 < a < -\pi$, and setting: $\theta = 2n\pi + a$ where n is an integer ≤ 0, we
get: $\theta/2 = (2n\pi + a)/2 = n\pi + a/2$, and $-3\pi/4 < a/2 < -\pi/2$.

And thus, if n is odd, it is in the first quadrant.
If however, n is even, it is in the third quadrant.

Suggestions or Solutions
To the **Problems 0 and 1** in the Example **2**

Two terminal rays are turning about the origin in the *x-y* plane. And both turn in the same direction. Suppose now, the angle made by one ray is θ, the angle made by the other ray is 6θ, and they are in a straight line, and indicate opposite directions, that is, the angle between the two is π. Then:

2.0.   Find the angle θ.   **2.1.   Find the angle θ assuming $0 < \theta < \pi/2$.**

Assuming first, the rays turn counterclockwise, and thus, setting: $6\theta - \theta = 2n\pi + \pi$ for n integer ≥ 0, we get: $5\theta = (2n + 1)\pi \Rightarrow \theta = (2n + 1)\pi/5$ for n integer ≥ 0.

And assuming next, the rays turn clockwise, and thus, setting: $\theta - 6\theta = 2n\pi + \pi$ for n integer ≥ 0, we get: $-5\theta = (2n + 1)\pi \Rightarrow \theta = -(2n + 1)\pi/5$ for n integer ≥ 0.

And next, assuming $0 < \theta < \pi/2$, we want to use $\theta = (2n + 1)\pi/5$ for n integer ≥ 0 since θ is positive, so we get:

$0 < (2n + 1)\pi/5 < \pi/2 \Rightarrow 0 < (2n + 1)\pi < 5\pi/2 \Rightarrow 0 < 2n + 1 < 5/2$
$\Rightarrow -1 < 2n < 3/2 \Rightarrow -1/2 < n < 3/4 \Rightarrow n = 0$ since n is an integer ≥ 0.

And thus, the angle θ is $\pi/5$ if $0 < \theta < \pi/2$.

If not quite sure of the idea behind the processes above, follow the steps below:

To begin with, can we just set: $6\theta - \theta = \pi$, since the angle between the two rays is π?

We don't know how many complete turns one ray has made before it made the angle θ. And also, we don't know how many complete turns the other ray has made before it made the angle 6θ.   How then, can we set the difference between the two angles made?

Unlike the previous example, the direction of turning does not really matter.

What matters is the difference between the two angles made, and it is π.

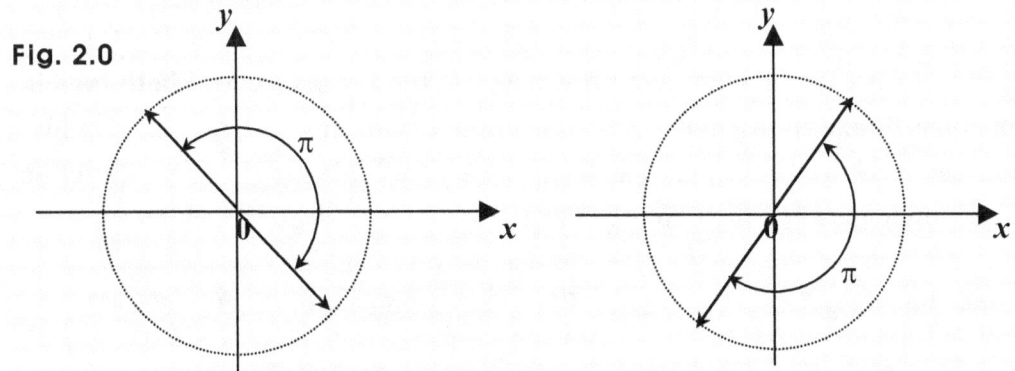

Fig. 2.0

And we don't know how many complete turns each ray made before each angle is made. What we know is though, the fact that making a complete turn, the angle made by each ray is 2π.

So assuming the rays turn counterclockwise, we can set the difference the way as follows: **6θ – θ = 2nπ + π** for *n* integer ≥ **0**.

Thus, we get: **5θ = (2n + 1)π ⇒ θ = (2n + 1)π/5** for *n* integer ≥ **0**.

And next, assuming the rays turn clockwise, we can set the difference the way as follows: **θ – 6θ = 2nπ + π** for *n* integer ≥ **0**.

Then, we get: **-5θ = (2n + 1)π ⇒ θ = -(2n + 1)π/5** for *n* integer ≥ **0**.

And next, we want to find the angle **θ** assuming **0 < θ < π/2**.

Then, the angle **θ** is positive, so we want to consider the case where the rays turn counterclockwise.

We have: **θ = (2n + 1)π/5**, and **0 < θ < π/2**.

So we get: **0 < (2n + 1)π/5 < π/2 ⇒ 0 < (2n + 1)π < 5π/2 ⇒ 0 < 2n + 1 < 5/2**
⇒ -1 < 2n < 3/2 ⇒ -1/2 < n < 3/4 ⇒ n = 0 since *n* is an integer ≥ **0**.

And thus, the angle **θ** is π/5 if **0 < θ < π/2**.

Suggestions or Solutions
To the **Problems 2 and 3** in the Example **2**

Two terminal rays are turning about the origin in the *x-y* plane. And both turn in the same direction. Suppose now, the angle made by one ray is θ, the angle made by the other ray is 6θ, and they are in a straight line, and indicate the same directions, that is, the angle between the two is 0. Then:

2.2. Find the angle θ.　　　　　　**2.3. Find the angle θ assuming $0 < \theta < \pi/2$.**

Assuming first, the rays turn counterclockwise, and thus, setting: $6\theta - \theta = 2n\pi$ for n integer ≥ 0, we get: $5\theta = 2n\pi \Rightarrow \theta = 2n\pi/5$ for n integer ≥ 0.

Assuming next, the rays turn clockwise, and thus, setting $\theta - 6\theta = 2n\pi$ for n integer ≥ 0, we get: $-5\theta = 2n\pi \Rightarrow \theta = -2n\pi/5$ for n integer ≥ 0.

And next, assuming $0 < \theta < \pi/2$, we want to use $\theta = 2n\pi/5$, since the angle θ is positive, so we get:

$0 < 2n\pi/5 < \pi/2 \Rightarrow 0 < 2n\pi < 5\pi/2 \Rightarrow 0 < 2n < 5/2 \Rightarrow 0 < n < 5/4$
$\Rightarrow n = 1$ since n is an integer ≥ 0.

And thus, the angle θ is $2\pi/5$ if $0 < \theta < \pi/2$.

If not quite sure of the idea behind the processes above, follow the steps below:

To begin with, can we just set: $6\theta - \theta = 0$, since the angle between the two rays is 0?

We don't know how many complete turns one ray has made before it made the angle θ. And also, we don't know how many complete turns the other ray has made before it made the angle 6θ. How then, can we set the difference between the two angles made?

As in the case of the previous example, what matters is the difference between the two angles made, and it is 0.

Fig. 2.1

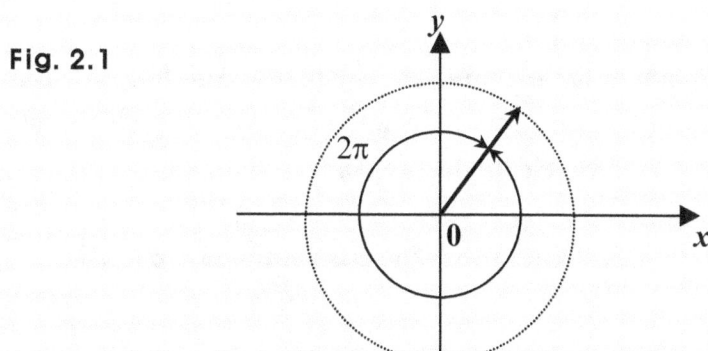

And since the angle between the two is 0, we can also say that the angle between the two is 2π. And we don't know how many complete turns each ray made before each angle is made. What we know is though, the fact that making a complete turn, the angle made by each ray is 2π.

So assuming the rays turn counterclockwise, we can set the difference the way as follows: $6\theta - \theta = 2n\pi$ for n integer ≥ 0.

Thus, we get: $5\theta = 2n\pi \Rightarrow \theta = 2n\pi/5$ for n integer ≥ 0.

And next, assuming the rays turn clockwise, we can set the difference the way as follows: $\theta - 6\theta = 2n\pi$ for n integer ≥ 0.

Then, we get: $-5\theta = 2n\pi \Rightarrow \theta = -2n\pi/5$ for n integer ≥ 0.

And next, we want to find the angle θ assuming $0 < \theta < \pi/2$.

Then, the angle θ is positive, so we want to consider the case where the rays turn counterclockwise.

We have: $\theta = 2n\pi/5$, and $0 < \theta < \pi/2$.

So we get: $0 < 2n\pi/5 < \pi/2 \Rightarrow 0 < 2n\pi < 5\pi/2 \Rightarrow 0 < 2n < 5/2 \Rightarrow 0 < n < 5/4 \Rightarrow n = 1$ since n is an integer ≥ 0.

And thus, the angle θ is $2\pi/5$ if $0 < \theta < \pi/2$.

Suggestions or Solutions
To the Problems 4 and 5 in the Example 2

Two terminal rays are turning about the origin in the *x-y* plane. And both turn in the same direction. Suppose now, the angle made by one ray is θ, the angle made by the other ray is 6θ, and they are symmetric about the *y*-axis. Then:

2.4.   Find the angle θ.   **2.5.**   Find the angle θ assuming 0 < θ < π/2.

Assuming first, the rays turn counterclockwise, and *n* is an integer ≥ 0, we can set:

θ + 6θ = 2*n*π + π or 2*n*π + 3π.

Meanwhile:
2*n*π + π = (2*n* + 1)π, and 2*n*π + 3π = 2*n*π + 2π + π = 2(*n* + 1)π + π = {2(*n* + 1) + 1}π.

And we know: **2*n* + 1** is an integer odd, and **{2(*n* + 1) + 1}** is an integer odd, too.
So we can set: **θ + 6θ = (2*n* + 1)π** for *n* integer ≥ 0.
And thus, we get: **7θ = (2*n* + 1)π ⇒ θ = (2*n* + 1)π/7.**

Assuming next, the rays turn clockwise, we can set: **θ + 6θ = -(2*n*π + π) or -(2*n*π + 3π).**

Meanwhile, **-(2*n*π + π) = -(2*n* + 1)π, and -(2*n*π + 3π) = -(2*n*π + 2π + π)**
= -{2(*n* + 1)π + π} = -{2(*n* + 1) + 1}π.

And we know: **2*n* + 1** is an integer odd, and so is **{2(*n* + 1) + 1}**.
So we can set: **θ + 6θ = -(2*n* + 1)π** for *n* integer ≥ 0.
And thus, we get: **7θ = -(2*n* + 1)π ⇒ θ = -(2*n* + 1)π/7.**

And next, assuming **0 < θ < π/2**, we want to use **θ = (2*n* + 1)π/7**, since the angle θ is positive, so we get: **0 < (2*n* + 1)π/7 < π/2 ⇒ 0 < (2*n* + 1)π < 7π/2 ⇒ 0 < 2*n* + 1 < 7/2**
⇒ -1 < 2*n* < 5/2 ⇒ -1/2 < *n* < 5/4 ⇒ *n* = 0 or 1 since *n* is an integer ≥ 0.

And thus, the angle θ is **π/7 or 3π/7 if 0 < θ < π/2.**

If not quite sure of the idea behind the processes above, follow the steps below:

We can have two cases where the two rays are symmetric about the *y*-axis.

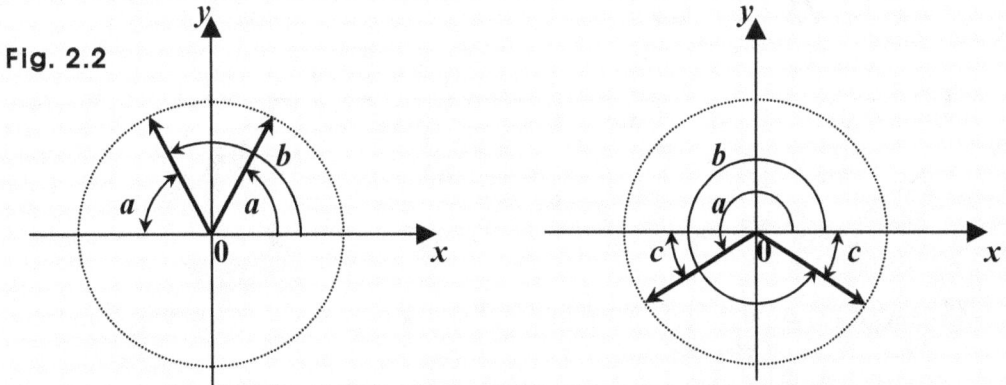

Fig. 2.2

Let's begin with the case where the rays turn counterclockwise.

Then first, in the graph on the left, we can see that $a + b = \pi$.

And next, in the graph on the right, we can see that $a = \pi + c$, and $b = 2\pi - c$, and thus, we get: $a + b = (\pi + c) + (2\pi - c) = 3\pi$.

And we don't know how many complete turns each of the rays has made.

So taking the sum of the two angles θ and 6θ, and assuming *n* is an integer ≥ 0, we can put it the way as follows: $\theta + 6\theta = 2n\pi + \pi$ or $2n\pi + 3\pi$.

However, we can put the sum the way below:
$2n\pi + \pi = (2n + 1)\pi$, and $2n\pi + 3\pi = 2n\pi + 2\pi + \pi = 2(n + 1)\pi + \pi = \{2(n + 1) + 1\}\pi$.

Thus, we get: $\theta + 6\theta = (2n + 1)\pi$ or $\{2(n + 1) + 1\}\pi$. So what?
We know: $2n + 1$ is an integer odd, and $\{2(n + 1) + 1\}$ is an integer odd, too.

So we can simply put the sum the way as follows: $\theta + 6\theta = (2n + 1)\pi$ for *n* integer ≥ 0.
And thus, we get: $7\theta = (2n + 1)\pi \Rightarrow \theta = (2n + 1)\pi/7$.

And let's next, move on to the case where the rays turn clockwise.
Then, to begin with, we can put some angles the way below:

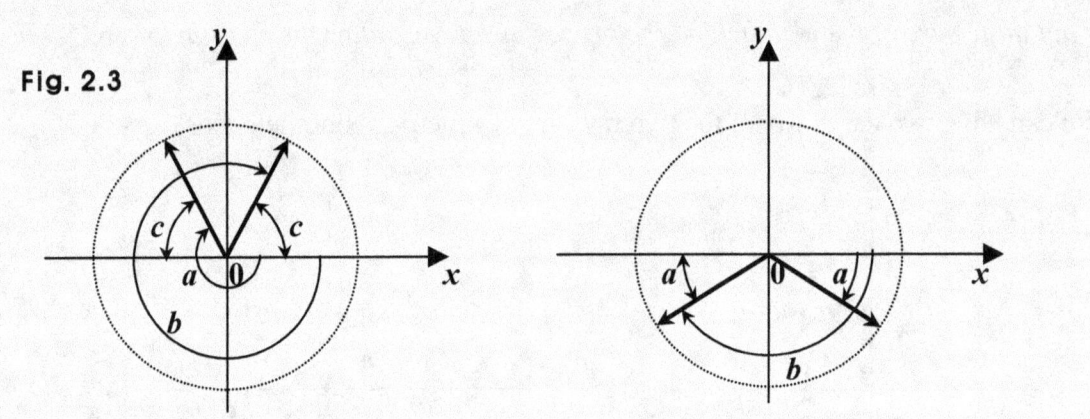

Fig. 2.3

Then first, in the graph on the right, we can see that $a + b = -\pi$.

And next, in the graph on the left, we can see that $a = -(\pi + c)$, and $b = -(2\pi - c)$, and thus, we get: $a + b = -(\pi + c) - (2\pi - c) = -3\pi$.

And we don't know how many complete turns each of the rays has made.

So taking the sum of the two angles θ and 6θ, and assuming n is an integer ≥ 0, we can put it the way as follows: $\theta + 6\theta = -(2n\pi + \pi)$ or $-(2n\pi + 3\pi)$.

And we can put the sum the way below:
$-(2n\pi + \pi) = -(2n + 1)\pi$, and $-(2n\pi + 3\pi) = -(2n\pi + 2\pi + \pi)$
$= -\{2(n + 1)\pi + \pi\} = -\{2(n + 1) + 1\}\pi$.

Thus, we get: $\theta + 6\theta = -(2n + 1)\pi$ or $-\{2(n + 1) + 1\}\pi$.

And we know: $2n + 1$ is an integer odd, and $\{2(n + 1) + 1\}$ is an integer odd, too.

So we can simply put the sum the way as follows: $\theta + 6\theta = -(2n + 1)\pi$ for n integer ≥ 0.

And thus, we get: $7\theta = -(2n + 1)\pi \Rightarrow \theta = -(2n + 1)\pi/7$.

And next, we want to find the angle θ assuming $0 < \theta < \pi/2$.

Then, the angle θ is positive, so we want to consider the case where the rays turn counterclockwise.

We have: $\theta = (2n + 1)\pi/7$, and $0 < \theta < \pi/2$.

So we get: $0 < (2n + 1)\pi/7 < \pi/2 \Rightarrow 0 < (2n + 1)\pi < 7\pi/2 \Rightarrow 0 < 2n + 1 < 7/2$

$\Rightarrow -1 < 2n < 5/2 \Rightarrow -1/2 < n < 5/4 \Rightarrow n = 0$ or 1 since n is an integer ≥ 0.

And thus, the angle θ is $\pi/7$ or $3\pi/7$ if $0 < \theta < \pi/2$.

In short:

Assuming first, the rays turn counterclockwise, and n is an integer ≥ 0, we can set:
$\theta + 6\theta = 2n\pi + \pi$ or $2n\pi + 3\pi$.

Meanwhile:
$2n\pi + \pi = (2n + 1)\pi$, and $2n\pi + 3\pi = 2n\pi + 2\pi + \pi = 2(n + 1)\pi + \pi = \{2(n + 1) + 1\}\pi$.

And we know: $2n + 1$ is an integer odd, and $\{2(n + 1) + 1\}$ is an integer odd, too.
So we can set: $\theta + 6\theta = (2n + 1)\pi$ for n integer ≥ 0.
And thus, we get: $7\theta = (2n + 1)\pi \Rightarrow \theta = (2n + 1)\pi/7$.

Assuming next, the rays turn clockwise, we can set: $\theta + 6\theta = -(2n\pi + \pi)$ or $-(2n\pi + 3\pi)$.

Meanwhile, $-(2n\pi + \pi) = -(2n + 1)\pi$, and $-(2n\pi + 3\pi) = -(2n\pi + 2\pi + \pi)$
$= -\{2(n + 1)\pi + \pi\} = -\{2(n + 1) + 1\}\pi$.

And we know: $2n + 1$ is an integer odd, and so is $\{2(n + 1) + 1\}$.
So we can set: $\theta + 6\theta = -(2n + 1)\pi$ for n integer ≥ 0.
And thus, we get: $7\theta = -(2n + 1)\pi \Rightarrow \theta = -(2n + 1)\pi/7$.

And next, assuming $0 < \theta < \pi/2$, we want to use $\theta = (2n + 1)\pi/7$, since the angle θ is positive, so we get:

$0 < (2n + 1)\pi/7 < \pi/2 \Rightarrow 0 < (2n + 1)\pi < 7\pi/2 \Rightarrow 0 < 2n + 1 < 7/2$

$\Rightarrow -1 < 2n < 5/2 \Rightarrow -1/2 < n < 5/4 \Rightarrow n = 0$ or 1 since n is an integer ≥ 0.

And thus, the angle θ is $\pi/7$ or $3\pi/7$ if $0 < \theta < \pi/2$.

60

Suggestions or Solutions
To the **Problems 6 and 7** in the Example **2**

Suppose two terminal rays are turning in the *x-y* plane, and they turn about the origin, of course. Suppose now, the angle made by one ray is θ, the angle made by the other ray is 6θ, and they are symmetric about the *x*-axis. Then:

2.6.   Find the angle θ.   **2.7.   Find the angle θ assuming $0 < \theta < \pi/2$.**

Assuming first, the rays turn counterclockwise, and *n* is an integer ≥ 0, we can set:

$$\theta + 6\theta = 2n\pi + 2\pi.$$

And we know: **$2n$** is an integer even, and so is **2**.

So we get: $7\theta = 2n\pi \Rightarrow \theta = 2n\pi/7$ for *n* integer ≥ 0.
Assuming next, the rays turn clockwise, we can set: $\theta + 6\theta = -(2n\pi + 2\pi)$.

And we know: **$2n$** is an integer even, and so is **2**.

So we get: $7\theta = -2n\pi \Rightarrow \theta = -2n\pi/7$ for *n* integer ≥ 0.

And thus, putting threads together, we get: $\theta = 2n\pi/7$ for *n* integer.

And next, assuming $0 < \theta < \pi/2$, we want to use $\theta = 2n\pi/7$ for *n* integer ≥ 0, since the angle θ is positive, so we get:

$$0 < 2n\pi/7 < \pi/2 \Rightarrow 0 < 2n\pi < 7\pi/2 \Rightarrow 0 < 2n < 7/2$$
$$\Rightarrow 0 < n < 7/4 \Rightarrow n = 1 \text{ since } n \text{ is an integer} \geq 0.$$

And thus, the angle θ is $2\pi/7$ if $0 < \theta < \pi/2$.

If not quite sure of the idea behind the processes above, follow the steps below:

As in the case of the previous example, we can have two cases where the two rays are symmetric about the *x*-axis.

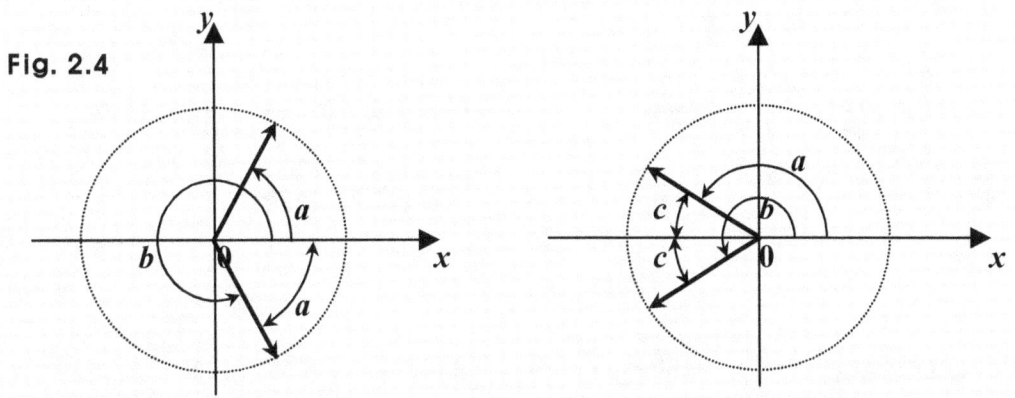

Fig. 2.4

Let's begin with the case where the rays turn counterclockwise.

Then first, in the graph on the left, we can see that $b = 2\pi - a$, so we get: $a + b = 2\pi$.

And next, in the graph on the right, we can see that $a = \pi - c$, and $b = \pi + c$, and thus, we get: $a + b = (\pi - c) + (\pi + c) = 2\pi$.

And we don't know how many complete turns each of the rays has made.

So taking the sum of the two angles θ and 6θ, and assuming *n* is an integer ≥ 0, we can put it the way as follows: $\theta + 6\theta = 2n\pi + 2\pi$.

And we know: $2n$ is an integer even, and 2 is an integer even, too.

So we can simply put the sum the way as follows: $\theta + 6\theta = 2n\pi$ for *n* integer ≥ 0.
And thus, we get: $7\theta = 2n\pi \Rightarrow \theta = 2n\pi/7$ for *n* integer ≥ 0.

And let's next, move on to the case where the rays turn clockwise.

Then, to begin with, we can put some angles the way below:

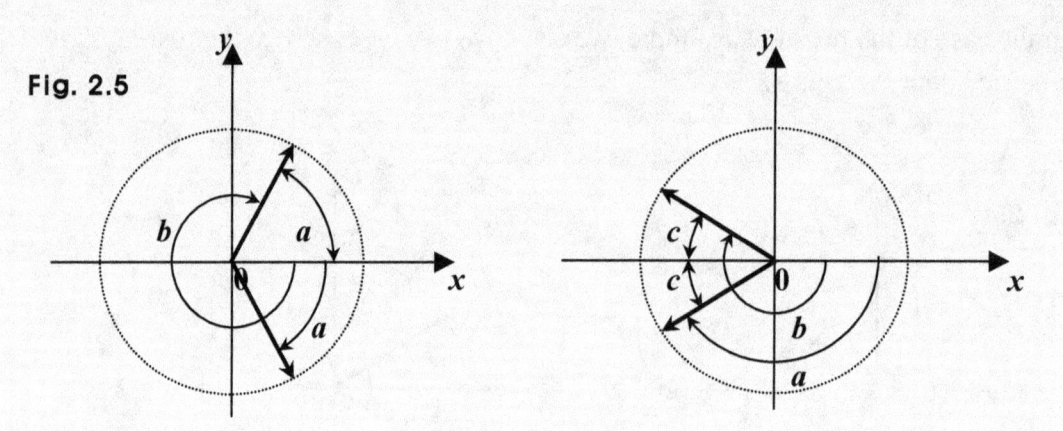

Fig. 2.5

Then first, in the graph on the right, we can see that $a + b = -2\pi$.

And next, in the graph on the left, assuming c is an angle positive, we can see that:

$a = -\pi + c$, and $b = -\pi - c$, and thus, we get: $a + b = -\pi + c - \pi - c = -2\pi$.

And we don't know how many complete turns each of the rays has made.

So taking the sum of the two angles θ and 6θ, and assuming n is an integer ≥ 0, we can put it the way as follows: $\theta + 6\theta = -2n\pi - 2\pi = -(2n\pi + 2\pi)$.

And we know: $2n$ is an integer even, and so is 2, of course.

So we can simply put the sum the way as follows: $\theta + 6\theta = 2n\pi$ for n integer ≤ 0.

And thus, we get: $7\theta = 2n\pi \Rightarrow \theta = 2n\pi/7$ for n integer ≤ 0.

Now, putting threads together, we have:

$\theta = 2n\pi/7$ for n integer ≥ 0 if the rays turn counterclockwise.

$\theta = 2n\pi/7$ for n integer ≤ 0 if the rays turn clockwise.

So either way, we can put it this way: $\theta = 2n\pi/7$ where n is an integer.

That's because the statement that n is an integer means that n can be positive, 0, or negative.

And if $n \geq 0$, the rays turn counterclockwise, and if $n < 0$, the rays turn clockwise.

What if $n = 0$ though?

If $n = 0$, we get: $\theta = 0$, and thus, $6\theta = 0$, too.

In that case, the two rays both are on the x-axis, which means technically, the two rays are symmetric about the x-axis.

And next, we want to find the angle θ assuming $0 < \theta < \pi/2$.

Then, the angle θ is positive, so we want to consider the case where the rays turn counterclockwise.

We have: $\theta = 2n\pi/7$, and $0 < \theta < \pi/2$. So we get:

$0 < 2n\pi/7 < \pi/2 \Rightarrow 0 < 2n\pi < 7\pi/2 \Rightarrow 0 < 2n < 7/2 \Rightarrow 0 < n < 7/4$
$\Rightarrow n = 1$ since n is an integer ≥ 0.

And thus, the angle θ is $2\pi/7$ if $0 < \theta < \pi/2$.

In short:

Assuming first, the rays turn counterclockwise, and n is an integer ≥ 0, we can set:

$\theta + 6\theta = 2n\pi + 2\pi.$

And we know: $2n$ is an integer even, and so is 2.

So we get: $7\theta = 2n\pi \Rightarrow \theta = 2n\pi/7$ for n integer ≥ 0.

Assuming next, the rays turn clockwise, we can set: $\theta + 6\theta = -(2n\pi + 2\pi)$.

And we know: $2n$ is an integer even, and so is 2.

So we get: $7\theta = -2n\pi \Rightarrow \theta = -2n\pi/7$ for n integer ≥ 0.

And thus, putting threads together, we get: $\theta = 2n\pi/7$ for n integer.

And next, assuming $0 < \theta < \pi/2$, we want to use $\theta = 2n\pi/7$ for n integer ≥ 0, since the angle θ is positive, so we get:

$0 < 2n\pi/7 < \pi/2 \Rightarrow 0 < 2n\pi < 7\pi/2 \Rightarrow 0 < 2n < 7/2$
$\Rightarrow 0 < n < 7/4 \Rightarrow n = 1$ since n is an integer ≥ 0.

And thus, the angle θ is $2\pi/7$ if $0 < \theta < \pi/2$.

2. Triangles and Trigonometry

So what is a triangle?

To begin with, tri means three, so literally, a triangle is made of three angles.

And next, a triangle is made of three sides, and is a polygon.
What is a polygon though?

It is a 2-D object closed, and is made of three or more line segments.

Usually though, we just call each line segment a side. And in basic math, we normally work with polygons said to be simple and convex. In each side in such a polygon, each endpoint is an endpoint of one of the other sides. And a line passing through such a polygon can cross up to two sides in the polygon, so it cannot cross three or more sides.

And among such polygons, we have triangles, quadrangles as rectangles, squares, rhombuses, parallelograms, and trapezoids, and others as pentagons, hexagons, etc.

So a triangle is a polygon made of three sides, in each of which, each endpoint is an endpoint of one of the other two sides.

And thus, triangles are the simplest polygons.

No matter what polygon it may be though, it can be said to be made of triangles. More specifically, it can be partitioned into triangles. And thus, triangles are not only the simplest but the most basic polygons, too.

And we can say also, a triangle is made of many other triangles, too, in each of which, one angle is a right angle, and thus, is 90°, which is π/2 rad. So they are called right triangles.

More specifically, no matter what triangle it may be, it can be partitioned into two right triangles at a time. And of course, we can partition a right triangle into two other right triangles at a time, also.

And thus, it can be said that every polygon is made of right triangles.

Fig. 0

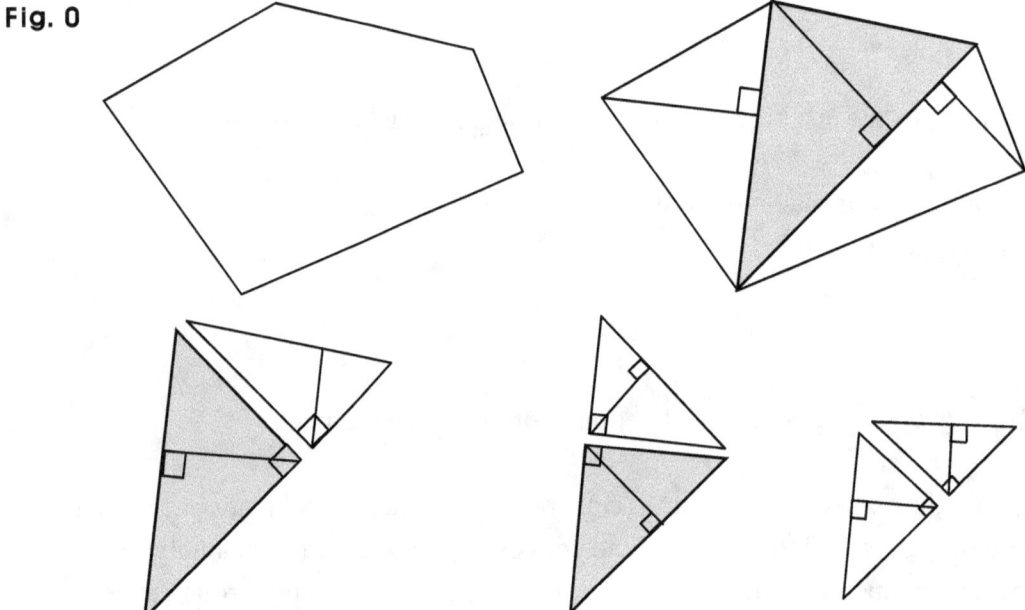

And we can keep partitioning a right triangle into two other right triangles.

So right triangles are fundamental triangles, and thus, are important. And they are very much so. It is often the case we can't do much without right triangles solving problems.

And we can do a lot using right triangles. Right triangles do much not only in geometry but in algebra, too. It is often vital that we use right triangles *right*. And thus, we want to know them very well, and use them very well, too.

Fig. 1

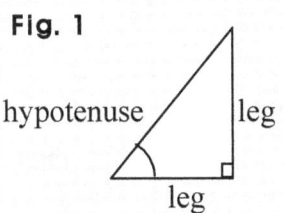

A right triangle is made of three sides, called a hypotenuse and two legs. And the two legs are perpendicular to each other, and thus, make 90°. So the hypotenuse is facing the angle 90°, and thus, is opposite of the right angle.

And the sum of the two angles adjacent to the hypotenuse is 90°, because the sum of all the three angles in a triangle is 180°.   How come it's 180° though?

To begin with, putting a triangle on a line called *x*, we can put it the way below:

Fig. 2

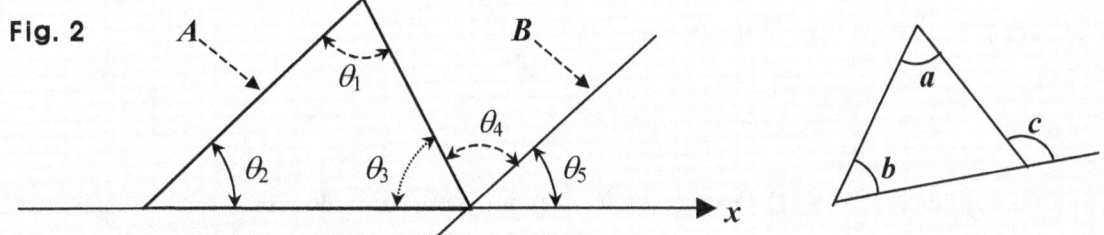

And assuming next, the line segment *A* is parallel to the line *B* above, we can say that:

The two angles θ_2 and θ_5 are corresponding angles, and thus, are equal, and the two angles θ_1 and θ_4 are alternate angles, and thus, are equal, too.

So we get: $\theta_1 + \theta_2 + \theta_3 = \theta_3 + \theta_4 + \theta_5$, which is 180°, that is, π rad.

Besides, we can say that in a triangle, the sum of two internal angles is the same as the external angle *supplementary* to the other internal angle. What do we mean by though, angles supplementary to each other?

If two angles are supplementary to each other, their sum is 180°. So in the figure above, we can say that: θ_3 is supplementary to the sum: $\theta_4 + \theta_5$, because: $\theta_3 + \theta_4 + \theta_5 = 180°$.

And thus, in the figure above, we get: $a + b = c$.

So if an angle is a *supplement* to another angle, the sum of the two angles is 180°.

By the way, if two angles are *complementary* to each other, their sum is 90°, a right angle. So if in a triangle, one angle is a *complement* to another, the other is a right angle, since the sum of the three angles is 180°. And thus, such a triangle is a right triangle.

• And next, from a right triangle, we can get an important tool, which is the distance formula, often called the Pythagorean theorem, too. What distance then, is it?

It is the distance between two points, and the distance is the length of the line segment connecting the two points. What two points though?

The two points can be in a 2-D space called a plane as the **x-y** plane, or in a 3-D space as the **x-y-z** space, where **x**, **y**, and **z** axes are perpendicular to each other.

So for instance, assuming first, **d** is the distance between two points (x_1, y_1) and (x_2, y_2), and using the theorem, we get:

• $d^2 = (x_2 - x_1)^2 + (y_2 - y_1)^2$, which equals: $(x_1 - x_2)^2 + (y_1 - y_2)^2$, too, of course.

And next, assuming **D** is the distance between two points (x_1, y_1, z_1) and (x_2, y_2, z_2), which are in a 3-D space, of course, and using the theorem again, we get:

$D^2 = d^2 + (z_2 - z_1)^2 = (x_2 - x_1)^2 + (y_2 - y_1)^2 + (z_2 - z_1)^2$.

And of course, given the two points (x_1, y_1, z_1) and (x_2, y_2, z_2), we can get the distance **D** at once the way below, too:

• $D^2 = (x_2 - x_1)^2 + (y_2 - y_1)^2 + (z_2 - z_1)^2$, which equals: $(x_1 - x_2)^2 + (y_1 - y_2)^2 + (z_1 - z_2)^2$.

And putting in graphs all the distance ideas above, we can put them the way below:

Fig. 3

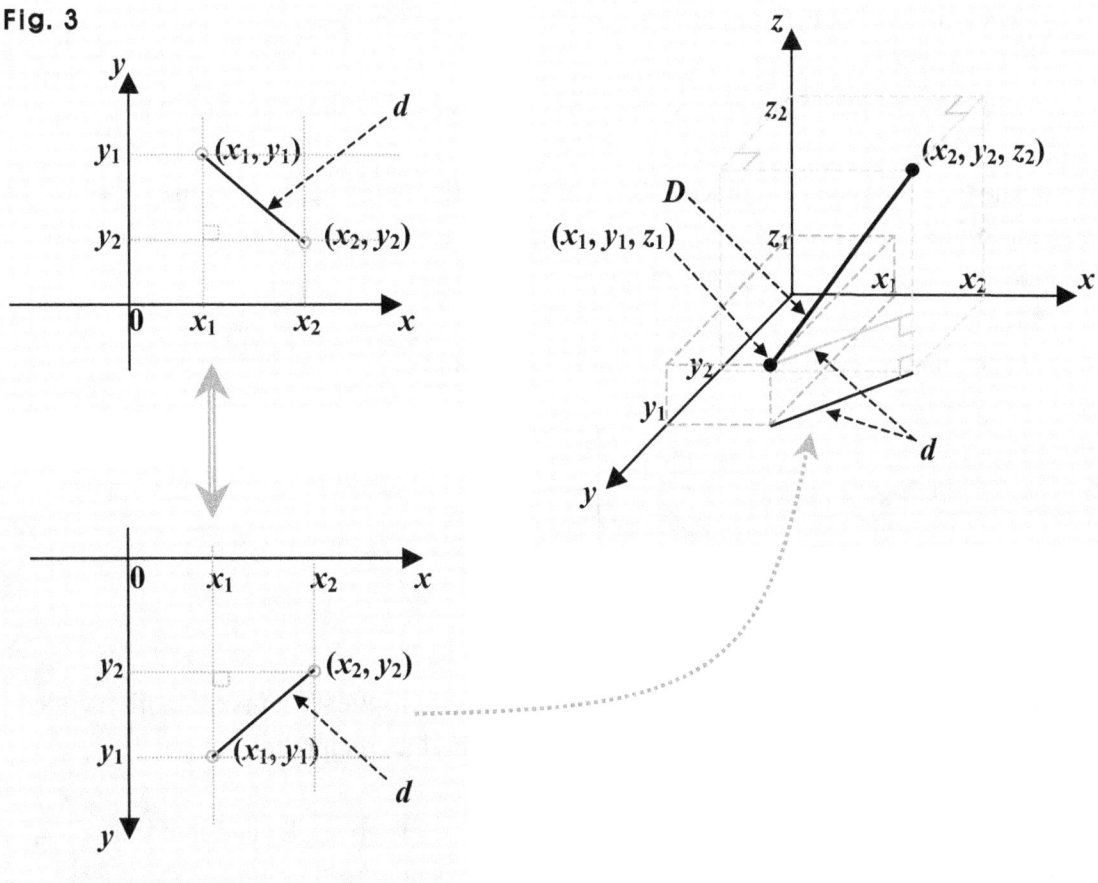

So is that all what a right triangle is about?

We have another important fact about a right triangle. And the fact is what this book is about, and is as follows:

• A right triangle is the place where a special geometry called trigonometry begins.

From a right triangle, we can get another important tool called a *trigonometric ratio*, called a *trig-ratio*, for short. We can get in fact, six of those in kinds.

Getting such a ratio, we use two of the three sides in a right triangle. So a trigonometric ratio is basically made of two sides in a right triangle.
What are the two sides though?

To begin with, we use different names for the legs of a right triangle.

And naming them, we consider an angle we refer to when getting a trig-ratio.

One of the two is called the adjacent side called briefly the adjacent, because it is adjacent to the angle stated above. And the other is called the opposite side called just the opposite, because it is opposite of the angle.

So in trigonometry, we put a right triangle the way below:

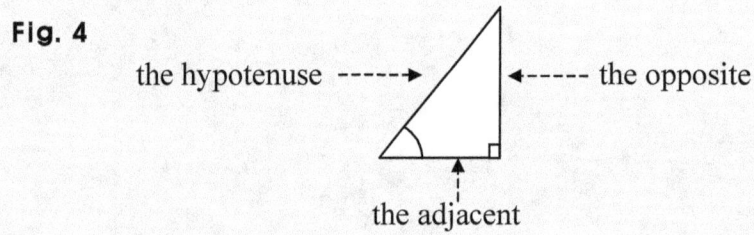

We have three basic trig-ratios, each of which has its multiplicative inverse, often called the reciprocal, too. So multiplying each of the three and its reciprocal, we get 1.

We want to note however, that the inverse of a trig-ratio is an angle, and that the multiplicative inverse or the reciprocal of a trig-ratio is another trig-ratio, which is a number. What are those three basic trig-ratios, then?

Getting a trigonometric ratio, we pick two of the three sides in a right triangle, and then, take a ratio between the two. We don't just pick two sides though, and do not simply take a ratio between the two.

Two of the three sides make an important angle in a right triangle. And we may want to call the important angle the governing angle. That's because such an angle governs or determines the ratios. What sides are the two though?

Assuming first, in a right triangle, H is the hypotenuse, A is the adjacent, and O is the opposite, we can put the right triangle the way below:

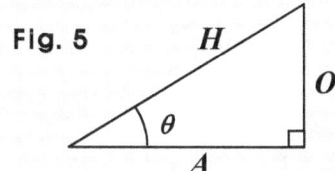

Fig. 5

Then, we call θ the governing angle, because we have chosen the side A to be the adjacent, or we have chosen the side O to be the opposite. So the adjacent is the side adjacent to the governing angle, and the opposite is the side opposite of or facing the governing angle.

And thus, the two sides are the adjacent and the hypotenuse. So in a right triangle, *the adjacent and the hypotenuse make the governing angle.*

And in return, *the governing angle determines every trig-ratio* in the right triangle.

Depending on the way we look at the right triangle though, either of the two sides other than the hypotenuse can be the adjacent. So using a wrong side as the adjacent, we get wrong ratios. Thus in short, *the wrong adjacent makes wrong ratios.*

So it is crucial to *choose the adjacent correctly.*

Now, we can get six trig-ratios in kind from a right triangle, and each is between two of the three sides. What then, are the six ratios?

Two are between the opposite and the hypotenuse, another two are between the adjacent and the hypotenuse, and the other two are between the opposite and the adjacent.

Of the six though, three are in fact, the reciprocals of the other three, which are called therefore, three basic trig-ratios. And more specifically, of the three basic ratios:

One is called *sine*, which is denoted by **sin**, and is the ratio of the opposite to the hypotenuse, that is, the opposite over the hypotenuse: the opposite / the hypotenuse.

Another is called *cosine*, which is denoted by **cos**, and is the ratio of the adjacent to the hypotenuse, that is, the adjacent over the hypotenuse: the adjacent / the hypotenuse.

And the other is called *tangent*, denoted by **tan**, which is the ratio of the opposite to the adjacent, that is, the opposite over the adjacent: the opposite / the adjacent. So the tangent can tell us the slope of the right triangle.

Suppose now, that the governing angle is θ, read as theta, which is the eighth letter of the Greek alphabet. Then, we put the three basic trig-ratios the way below:

• To begin with, <u>the sine of the governing angle θ</u> is: **sin** θ, and is the ratio of the opposite to the hypotenuse, so **sin** θ is: <u>the opposite over the hypotenuse</u>.

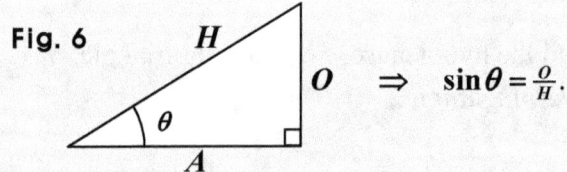

Fig. 6 $\quad H \quad O \quad \Rightarrow \quad \sin\theta = \frac{O}{H}.$

And we read **sin** θ as: <u>sine of θ</u> or just <u>sine θ</u>, for short. So for instance, we read **sin 30°** as <u>sine of 30°</u> or just <u>sine 30°</u>, which equals: **sin π/6**, since 30° is **π/6** in radian.

• Next, <u>the cosine of the governing angle θ</u> is **cos** θ, and is the ratio of the adjacent to the hypotenuse, so **cos** θ is: <u>the adjacent over the hypotenuse</u>.

Fig. 7 $\quad H \quad O \quad \Rightarrow \quad \cos\theta = \frac{A}{H}.$

And we read **cos** θ as: <u>cosine of θ</u> or just <u>cosine θ</u>, for short. So for instance:
We read **cos 45°** as <u>cosine of 45°</u> or just <u>cosine 45°</u>, which equals: **cos π/4**.

• And next, <u>the tangent of the governing angle θ</u> is **tan** θ, and is the ratio of the opposite to the adjacent, so **tan** θ is: <u>the opposite over the adjacent</u>.

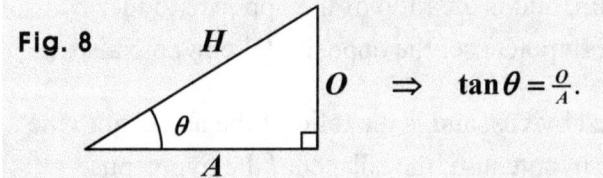

Fig. 8 $\quad H \quad O \quad \Rightarrow \quad \tan\theta = \frac{O}{A}.$

And we read **tan** θ as: <u>tangent of θ</u> or just <u>tangent θ</u>, for short.

So for instance, we read **tan 60°** as <u>tangent of 60°</u> or just <u>tan of 60°</u>, which is equal to: **tan π/3**. And we can call the tangent the <u>slope of the hypotenuse</u>, because the tangent is: the opposite over the adjacent.

And thus, putting threads together, and assuming X is the adjacent in the right triangle below, we have to use θ as the governing angle, and can put the three basic trig-ratios the way as follows:

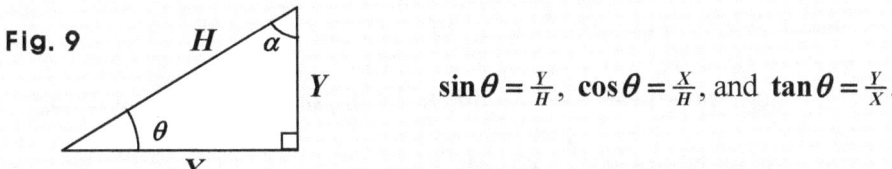

Fig. 9 $\sin\theta = \frac{Y}{H}$, $\cos\theta = \frac{X}{H}$, and $\tan\theta = \frac{Y}{X}$.

And for another instance, assuming α is the governing angle in the right triangle below, we use Y as the adjacent, and can put the three basic trig-ratios the way below:

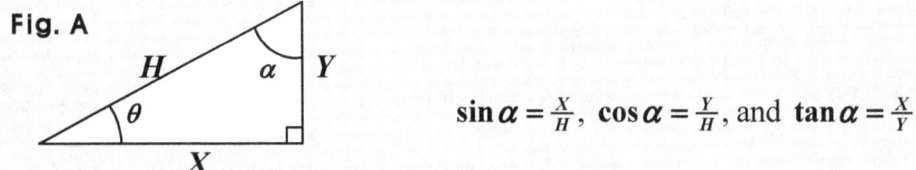

Fig. A $\sin\alpha = \frac{X}{H}$, $\cos\alpha = \frac{Y}{H}$, and $\tan\alpha = \frac{X}{Y}$.

And we have three popular governing angles, which are 30°, 45°, and 60°. That's because, we can easily get the trig-ratios for those angles using two triangles isosceles.

One is a regular (equilateral) triangle, where every angle is 60°, and the other is a right triangle isosceles, where two of the three angles are equal, and thus, are 45° each.

So to begin with, cutting in half a regular triangle, we can get a right triangle as below:

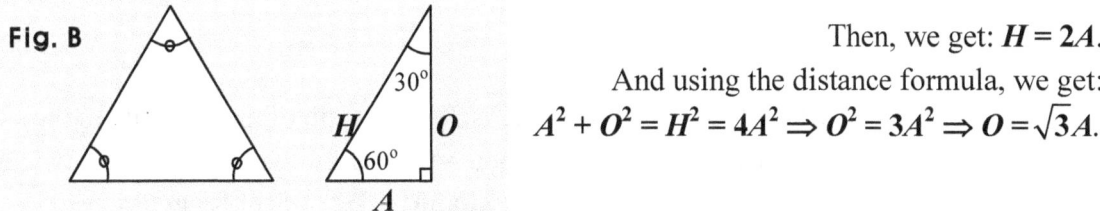

Fig. B   Then, we get: $H = 2A$.
And using the distance formula, we get:
$$A^2 + O^2 = H^2 = 4A^2 \Rightarrow O^2 = 3A^2 \Rightarrow O = \sqrt{3}A.$$

So we get:

Fig. C

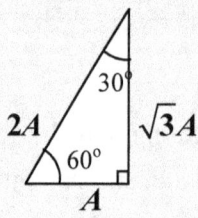

And thus, we can readily get:

sin 60° = the opposite / the hypotenuse = $\frac{\sqrt{3}}{2}$. **sin 30°** = the opposite / the hypotenuse = $\frac{1}{2}$.

cos 60° = the adjacent / the hypotenuse = $\frac{1}{2}$. **cos 30°** = the adjacent / the hypotenuse = $\frac{\sqrt{3}}{2}$.

tan 60° = the opposite / the adjacent = $\frac{\sqrt{3}}{1}$. **tan 30°** = the opposite / the adjacent = $\frac{1}{\sqrt{3}} = \frac{\sqrt{3}}{3}$.

And next, assuming A is the adjacent of a right triangle isosceles, we can put the triangle the way below:

Fig. D

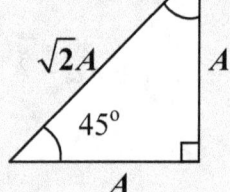

So we get:

sin 45° = the opposite / the hypotenuse = $\frac{1}{\sqrt{2}} = \frac{\sqrt{2}}{2}$.

cos 45° = the adjacent / the hypotenuse = $\frac{1}{\sqrt{2}} = \frac{\sqrt{2}}{2}$.

tan 45° = the opposite / the adjacent = $\frac{1}{1} = 1$.

Is it the case though, for instance, we get: **sin 30° = 1/2** in every right triangle where the governing angle is 30°?

Yes, it is. So for another instance, we get: **cos 60° = 1/2** in every right triangle where the governing angle is 60°. And the same is true for all the other trig-ratios, too. How come?

In *every* right triangle where the governing angle is 30°, the ratio of the opposite to the hypotenuse is 1/2. So the opposite is half the hypotenuse, that is, the hypotenuse is twice the opposite. And in this case, we put it this way: **sin 30° = 1/2.**   Still foggy?

If governing angles are the same, all the right triangles are said to be similar. That's because in all those triangles, all the corresponding angles are the same.

Suppose for instance, in a triangle, the three angles are 35°, 75°, and 80°, and in another triangle, the three angles are 35°, 75°, and 80°, too.

Then, the two triangles are called similar triangles.

And we can put (infinitely) many similar right triangles the way below:

Fig. E

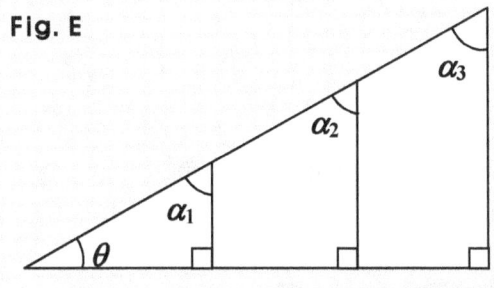

All the right triangles on the left are said to be

similar to each other.
And in triangles similar to each other,
all **corresponding angles** are the **same**.
That is, $\alpha_1 = \alpha_2 = \alpha_3$,
because α_1, α_2, and α_3 are corresponding angles.

In each right triangle above, assuming θ is the angle between the adjacent and the hypotenuse, we take θ as the governing angle. So in all those triangles, **sin θ** is the same.

And thus, *no matter what* right triangle it may be, if the governing angle is 30°, the ratio of the opposite to the hypotenuse is: **sin 30°**, which is 1/2.

So for instance, if in a right triangle where the governing angle is 30°, the hypotenuse is 1, the opposite is 1/2. And if in another right triangle where the governing angle is 30°, the hypotenuse is 2, the opposite is 1.

What then, can we simply call the trig-ratio **sin 30°**?

We can call it the opposite, which is in the right triangle where the governing angle is 30°, and the hypotenuse is 1.

So we can simplify the concept of the trig-ratio called the sine the way below:

- The *sine* is the *opposite* if the hypotenuse is 1.

What then, about the **sin** θ?

It is the opposite, if the governing angle is θ, and the hypotenuse is 1. So in short, we can put it the way below:

- The **sin** θ is the *opposite* if the **hypotenuse** is **1**.

And the same is true, too, for the **cos** θ and **tan** θ. So in short, we can put them the way below:

- The **cos** θ is the *adjacent* if the **hypotenuse** is **1**.

- The **tan** θ is the *opposite* if the **adjacent** is **1**.

- What then, about the reciprocal of each?

The reciprocal of the sine is called the *cosecant*, which is denoted by **csc**, and is the ratio of the hypotenuse to the opposite, that is, the hypotenuse over the opposite, since the sine is the ratio of the opposite to the hypotenuse: the opposite over the hypotenuse.

Fig. F H O $\Rightarrow$ $\csc\theta = \frac{H}{O}$, since it is the reciprocal of $\sin\theta = \frac{O}{H}$.

θ

A

And we read **csc** θ as: <u>cosecant of θ</u> or just <u>cosecant θ</u> or simply, <u>cosec θ</u>, for short.

Next, the reciprocal of the cosine is called the *secant*, which is denoted by **sec**, and is the ratio of the hypotenuse to the adjacent, that is, the hypotenuse over the adjacent, since the cosine is the ratio of the adjacent to the hypotenuse: the adjacent over the hypotenuse.

Fig. G

$O \Rightarrow \mathbf{sec}\,\theta = \frac{H}{A}$, since it is the reciprocal of $\mathbf{cos}\,\theta = \frac{A}{H}$.

And we read **sec** θ as: <u>secant of θ</u> or just <u>secant θ</u> or simply, <u>sec θ</u>, for short.

And the reciprocal of the tangent is called the *cotangent*, which is denoted by **cot**, and is the ratio of the adjacent to the opposite, that is, the adjacent over the opposite, since the tangent is the ratio of the opposite to the adjacent: the opposite over the adjacent.

Fig. H

$O \Rightarrow \mathbf{cot}\,\theta = \frac{A}{O}$, since it is the reciprocal of $\mathbf{tan}\,\theta = \frac{O}{A}$.

And we read **cot** θ as: <u>cotangent of θ</u> or just <u>cotan of θ</u> or simply, <u>cotan θ</u>, for short.

So for instance, what do we mean by **csc 30°**, which is 2?

It is a trig-ratio, and thus, is a ratio, which indicates the *relative* amount of the side called the *hypotenuse* with respect to the amount of the side called the *opposite* in *every* right triangle where the *governing angle* is 30°, which is the angle between the hypotenuse and the side called the adjacent.

And the *relative* amount stated above does not mean the actual length of the side called the hypotenuse, but is a ratio called a trig-ratio.

And in this case, the ratio is **csc 30°**, which is 2, which is the ratio of the *hypotenuse* (with respect) to the opposite.

So for instance, if in a right triangle where the governing angle is 30°, the *opposite* is 1, the *hypotenuse* is 2. And if in another right triangle where the governing angle is 30°, the opposite is 2, the hypotenuse is 4.

What then, can we simply call the trig-ratio **csc 30°**?

We can call it the *hypotenuse* if the governing angle is 30°, and the opposite is 1.

So we can simplify the concept of the trig-ratio called the cosecant the way below:

• The *cosecant* is the *hypotenuse* in the right triangle where the opposite is 1.

What then, about the **csc θ**?

It is the *hypotenuse* if the governing angle is θ, and the opposite is 1.
So in short, we can put it the way as follows:

   • The **csc θ** is the *hypotenuse* if the opposite is 1.

And the same is true, too, for the **sec θ** and the **cot θ**. So in short, we can put them the way below:

   • The **sec θ** is the *hypotenuse* if the adjacent is 1.

   • The **cot θ** is the *adjacent* if the *opposite* is 1.

And of course, the angle θ is the governing angle, which is the angle between the hypotenuse and the adjacent.

And thus, putting threads together, we have:

- The **sin** θ is the *opposite* if the hypotenuse is 1.
- The **cos** θ is the *adjacent* if the hypotenuse is 1.
- The **tan** θ is the *opposite* if the *adjacent* is 1.

- The **csc** θ is the *hypotenuse* if the opposite is 1.
- The **sec** θ is the *hypotenuse* if the adjacent is 1.
- The **cot** θ is the *adjacent* if the *opposite* is 1.

- What then, about the inverses?

The multiplicative inverses of the basic trig-ratios are the reciprocals of the basic ones.

So the multiplicative inverse of **sin** θ is **csc** θ.

Saying however, just the inverse of a trig-ratio, we mean an angle.

Thus for instance, the inverse of **sin** θ is θ, and the inverse of **cos 60°** is **60°**.

And the inverse of the sine is denoted by **sin**$^{-1}$, which is however, read as the *arc sine*.

Why arc though?

What can we mean by an arc?

It can indicate an angle, of course.

So the *arc sine* is the angle, the sine of which is the ratio.

Thus for instance, we know **sin 30°** is $\frac{1}{2}$. So **sin**$^{-1}\frac{1}{2}$ is **30°**.

And **sin**$^{-1}\frac{1}{2}$ is read as the arc sine of $\frac{1}{2}$. So the arc sine of $\frac{1}{2}$ is 30°, which is $\pi/6$.

And of course, the same is true for the other trig-ratios, too.

So the inverse of the cosine is denoted by **cos**$^{-1}$, read as the *arc cosine*, and the inverse of the tangent is denoted by **tan**$^{-1}$, read as the *arc tangent* or just the *arc tan*.

So for instance, the arc cosine of $\frac{1}{2}$ is **cos**$^{-1}\frac{1}{2}$, which is 60°, and the arc tan of 1 is **tan**$^{-1}$ **1**, which is 45°. And the same is true, too, for the reciprocals: **csc, sec,** and **cot**.

So the *arc cosecant* is **csc**$^{-1}$, the *arc secant* is **sec**$^{-1}$, and the *arc cotangent* is **cot**$^{-1}$. And each of them is an angle, too, of course.

Thus for instance, we can have:

csc^{-1} **2** = 30°, which is $\pi/6$, since **sin 30°** = 1/2 so **csc 30°** = 2.

sec^{-1} **2** = 60°, which is $\pi/3$, since **cos 60°** = 1/2 so **sec 60°** = 2.

cot$^{-1}\frac{1}{\sqrt{3}}$ = 60°, which is $\frac{\pi}{3}$, since **tan 60°** = $\sqrt{3}$ so **cot 60°** = $\frac{1}{\sqrt{3}}$, which is $\frac{\sqrt{3}}{3}$.

An Example in Triangles and Trigonometry

Find the value of the cosine of 15°, that is, **cos 15°**, algebraically, of course.

By the way, if an angle is very small, and is in radian, we can get each trig-ratio of the angle the way below, too:

(Note that the formulas below are not covered in high school math.)

$$\sin x = \sum_{n=0}^{\infty} (-1)^n \frac{x^{2n+1}}{(2n+1)!}$$

$$\cos x = \sum_{n=0}^{\infty} (-1)^n \frac{x^{2n}}{2n!}$$

$$\tan x = \sin x / \cos x$$

For instance, $\sin 0.1 = \sum_{n=0}^{\infty} (-1)^n \frac{0.1^{2n+1}}{(2n+1)!}$

$= \{(-1)^0 0.1^1\}/1! + \{(-1)^1 0.1^3\}/3! + \{(-1)^2 0.1^5\}/5! + \{(-1)^3 0.1^7\}/7! + \dots$

$= (1 \times 0.1)/1 - 0.1^3/(1 \cdot 2 \cdot 3) + 0.1^5/(1 \cdot 2 \cdot 3 \cdot 4 \cdot 5) - 0.1^7/(1 \cdot 2 \cdot 3 \cdot 4 \cdot 5 \cdot 6 \cdot 7) + \dots$

$= 0.1/1 - 0.1^3/6 + 0.1^5/120 - 0.1^7/720 + \dots$

$= 0.1 - 0.001/6 + 0.00001/120 - 0.0000001/720 + \dots$

Suggestions or Solutions
To the Example in Triangles and Trigonometry

Find the value of the cosine of 15°, that is, cos 15°.

We have: $\cos 30° = \frac{\sqrt{3}}{2}$, $\cos 45° = \frac{\sqrt{2}}{2}$, and $\cos 60° = \frac{1}{2}$.

And getting the ratios above, we use two right triangles, one is half a regular triangle, and the other is a right triangle isosceles.
How then, can we get the ratio, **cos** 15°?

We know 15° is the governing angle. So?

So using a right triangle where an angle is 15°, we should be able to get the value of the trig-ratio, **cos** 15°. And thus, forming such a triangle, we can get:

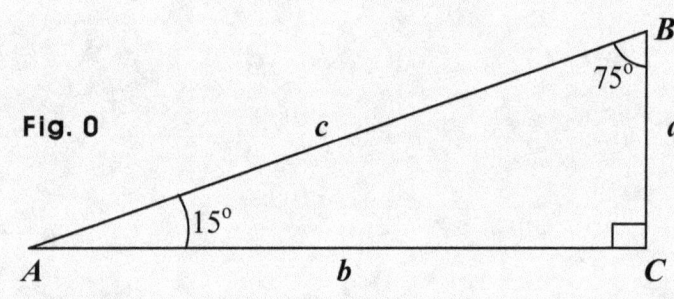

Fig. 0

We know that the cosine is:
the adjacent over the hypotenuse.

So we can set: **cos 15°** = *b*/*c*. But we don't know the values of *b* and *c*.

We know however: $\cos 30° = \frac{\sqrt{3}}{2}$, $\cos 45° = \frac{\sqrt{2}}{2}$, and $\cos 60° = \frac{1}{2}$.

So we may want to take advantage of those ratios. How though?

We can notice that we can form inside the triangle above a right triangle where two of the angles are 30° and 60°. And thus, forming it, we can get:

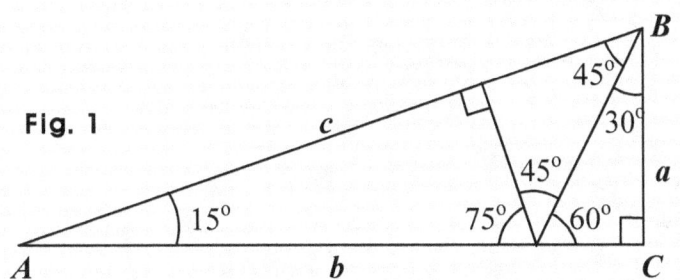

Fig. 1

We don't really get anything
new though.

So we may want to try again, forming another right triangle, where two of the angles are 30° and 60°, of course.

And thus, forming it a bit differently, we can get the one below:

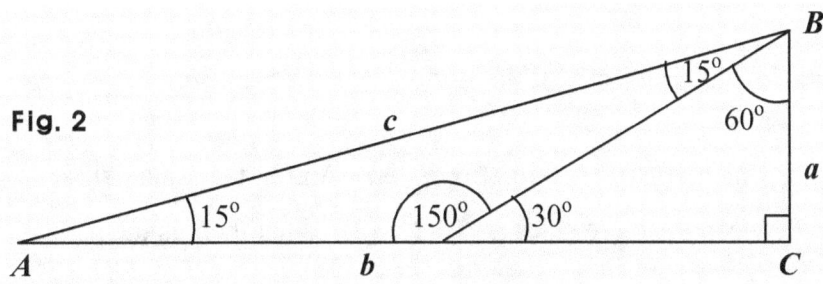

Fig. 2

Now, we can see something useful. What then, is it?

It is an isosceles triangle, where two of the three angles are 15° each.

So we may want to put labels a bit differently, indicating angles and sides. And we can put them the way below:

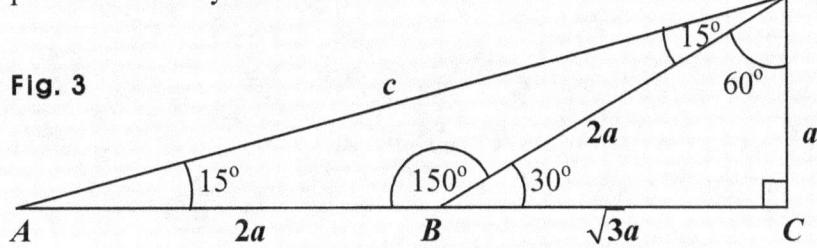

Fig. 3

How do we get though, $BC = \sqrt{3}a$, and $BD = 2a$?

We know the triangle **BCD** is half the regular triangle where each side is **2a**.

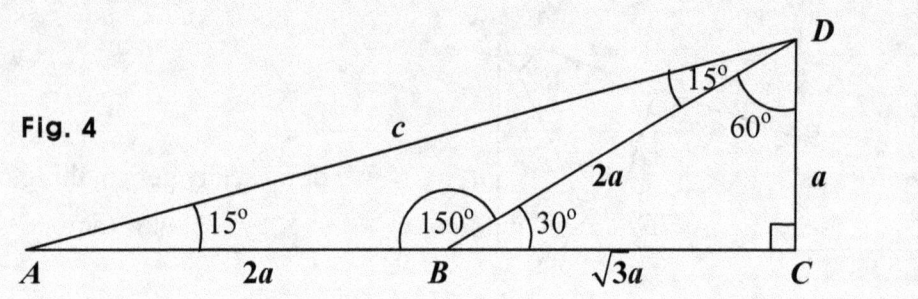

Fig. 4

So **BD** = **2a**. And we can get the length of **BC** using the distance formula, often called the Pythagorean theorem, too.

Assuming thus, *x* is the length of **BC**, and using the formula, we get:

$$(BD)^2 = (BC)^2 + (CD)^2 \Rightarrow 4a^2 = x^2 + a^2 \Rightarrow x^2 = 4a^2 - a^2 = 3a^2 \Rightarrow x = \pm\sqrt{3}a.$$

And we know *x* is a length, which is ≥ 0. So we get: $x = \sqrt{3}a$. And in fact, in a half a regular triangle, the ratio between the three sides is: $1 : 2 : \sqrt{3}$. What then, is the next?

We can get the length of the side **AC**, which is the adjacent if the angle *A* is the governing angle. And the length is: $2a + \sqrt{3}a$, which is the adjacent in the triangle **ACD**.

Now, we know the cosine is: the adjacent over the hypotenuse.
So what do we want to find?

It is the hypotenuse, which is *c* in the right triangle **ACD** above.
How then, can we get the value of *c*?

We can get it using the distance formula. too. So using the formula again, we can get:

$$(AD)^2 = (AC)^2 + (CD)^2 \Rightarrow c^2 = (2a + \sqrt{3}a)^2 + a^2 = 4a^2 + 4\sqrt{3}\,a^2 + 3a^2 + a^2 = (8 + 4\sqrt{3}\,)a^2.$$

So we get: $c^2 = (8 + 4\sqrt{3})a^2$. And thus, we get: $c = a\sqrt{8 + 4\sqrt{3}}$, since c is a length.

Doing some algebra though, we can put c the way below, too:

$$8 + 4\sqrt{3} = 8 + 2 \cdot 2\sqrt{3} = 8 + 2\sqrt{4}\sqrt{3} = 8 + 2\sqrt{12} = 8 + 2\sqrt{2 \cdot 6} = 8 + 2\sqrt{2}\sqrt{6}$$

$$= 2 + 6 + 2\sqrt{2}\sqrt{6} = (\sqrt{2} + \sqrt{6})^2 \Rightarrow 8 + 4\sqrt{3} = (\sqrt{2} + \sqrt{6})^2 \Rightarrow c = (\sqrt{2} + \sqrt{6})a.$$

Now, we know that the cosine is: the adjacent over the hypotenuse.

And the adjacent is: $2a + \sqrt{3}a$, which is $(2 + \sqrt{3})a$, and the hypotenuse is: $(\sqrt{2} + \sqrt{6})a$.

So we can put the cosine of $15°$, that is, **cos 15°** the way below:

$$\mathbf{cos\ 15°} = (2 + \sqrt{3})a / (\sqrt{2} + \sqrt{6})a = \frac{2 + \sqrt{3}}{\sqrt{2} + \sqrt{6}} = \frac{(2 + \sqrt{3})}{\sqrt{2}(1 + \sqrt{3})} = \frac{(2 + \sqrt{3})(\sqrt{3} - 1)}{\sqrt{2}(1 + \sqrt{3})(\sqrt{3} - 1)}$$

$$= \frac{1 + \sqrt{3}}{2\sqrt{2}} = \frac{\sqrt{2} + \sqrt{6}}{4} \Rightarrow \cos 15° = \frac{\sqrt{2} + \sqrt{6}}{4}.$$

Isn't there any other way though, to find the length of the line segment **AD**?

We can get it using a formula called the cosine rule, too, which will be covered in the section called **The Cosine Rule**.

And assuming θ is the angle **ABD** in the triangle **ABD** above, and using the cosine rule, we get: $(AD)^2 = (AB)^2 + (BD)^2 - 2(AB)(BD)$ **cos** θ.

So we get: $c^2 = 4a^2 + 4a^2 - 2 \cdot 2a \cdot 2a \cdot \cos 150° = 8a^2 + 2 \cdot 4a^2 \cdot \cos 30° = (8 + 4\sqrt{3})a^2$.

3. **What to do with Trig—Ratios 1**

That is, what can we do with **sin θ**, **cos θ**, etc.?

Taking care of objects that have to do with angles, we often use trig-ratios.

A *vector* is a good example.
A vector is an object with an angle.
So taking care of vectors, we often need to work with trig-ratios.

What do we mean by though, an object with an angle?

A vector is an object that has a magnitude and an angle.
And also, it is often said that a vector has an amount and a direction.

• So a vector has not only a *size* but an *angle*, too.

And among vectors, we have forces, velocities, etc. So for instance, a force has its magnitude and its angle (or direction).   How then, do we describe a particular vector?

We know that an arrow has its length and an angle or a direction.
So we can say that an arrow has its magnitude and an angle.
Describing thus, a vector particular, we usually use an arrow, the length of which is the magnitude of the vector. And the angle of the arrow is the angle that the vector has.

And also, we can put a vector this way, too:

> • A vector is said to have its *components*, *perpendicular* to each other.

If for instance, a vector is in a 2-D space as the *x-y* plane, it can be said to have two components. One is called the horizontal component, which is parallel to the *x*-axis, and the other is called the vertical component, which is parallel to the *y*-axis.

> • And getting each component, we can use one of trig-ratios as **sin *θ*, cos *θ*,** etc.

For what then, can we use such a ratio?   That is, what can we do with such a ratio?

We can use it finding such a component stated above.

Showing a vector, we often put it on a coordinate plane as the *x-y* plane.

And describing the vector, we use an arrow to specify the vector's size by the length of the arrow, and specify the vector's angle by the angle of the arrow.

And usually, specifying the angle, we specify the angle the arrow makes with the *x*-axis.

So the angle between the arrow and the *x*-axis is the angle of the vector. And in fact, just saying the angle of a vector, we mean the angle between the vector and the *x*-axis.

Assuming now, that *θ* is the angle between the vector and the *x*-axis, we can get the vertical component using a trig-ratio called **sin *θ*,** and the horizontal component using a trig-ratio called **cos *θ*.**

Also, using a trig-ratio called **tan *θ*,** we can specify the slope of the vector, because the **tan *θ*** is the slope of the arrow, which is a line segment.

So for instance, looking at each of the right triangles we have made earlier studying the trig-ratios, we can notice that the *hypotenuse* can be taken as (the size of) a vector.   So?

So we can take the adjacent as the horizontal component, and can take the opposite as the vertical component.   How then, can we get the components?

(Note that normally, specifying a vector, we put a small arrow on top of a letter saying the name of the vector as $\overrightarrow{H}$. Just for simplicity however, it is omitted in this book.)

Assuming H is a vector, and its angle is θ, we can put it in the x-y plane the way below:

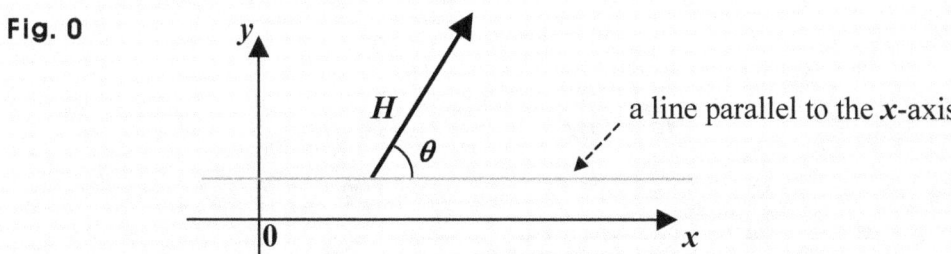

Fig. 0

a line parallel to the x-axis

Then, finding the components of H, we can put H in a right triangle the way below:

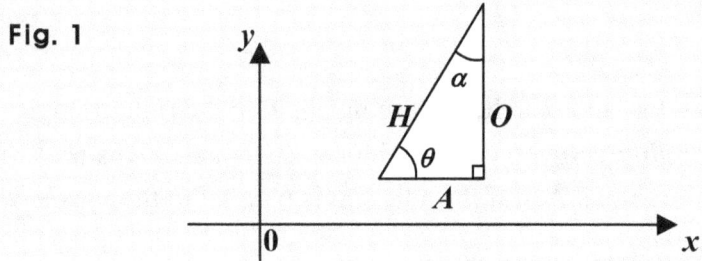

Fig. 1

Then, H is the hypotenuse, and A is the horizontal component. And taking the angle θ as the governing angle, we can say that A is the adjacent.

How then, can we get the adjacent A, that is, the horizontal component?

We know that the cosine is: the adjacent over the hypotenuse. So we get: $\cos \theta = \frac{A}{H}$.

And also, we can have: $H \cdot \frac{A}{H} = A$, which is the adjacent, which is in this case, the horizontal component.   So we get: $H \cdot \cos \theta = H \cdot \frac{A}{H} = A$.

So *multiplying* the hypotenuse H by **cos θ**, we get: $H \cdot \cos \theta = H \cdot \frac{A}{H} = A$, the adjacent.

And thus, multiplying the hypotenuse by the cosine, we get the adjacent.

Assuming for instance, the angle θ is 60°, and putting H in the *x-y* plane, we can get:

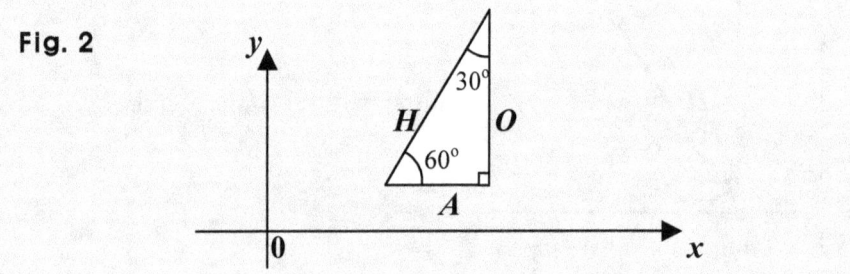

Fig. 2

Then, the governing angle is **60°**, so we get: $H \cdot \cos 60^\circ = H \cdot \frac{1}{2} = A$, since **cos 60° = 1/2**. So we get: $A = \frac{H}{2}$.

So in general, assuming V is a vector, and θ is the governing angle in the right triangle where V is the hypotenuse, and *multiplying* the vector V by **cos θ**, we get: $V \cdot \cos \theta$, which is the adjacent in the right triangle. And if the adjacent is parallel to the *x*-axis, we call it the horizontal component.

And by the same token, *applying* **sin θ** to the hypotenuse H, we get:

$H \cdot \sin \theta = H \cdot \frac{O}{H} = O$, which the opposite in the right triangle above, and is in this case, the vertical component of the vector H.

So of course, in general, assuming V is a vector, and θ is the governing angle in the right triangle where V is the hypotenuse, and *multiplying* **sin θ** to the vector V, we get: $V \cdot \sin \theta$, which is the opposite in the right triangle. And if the opposite is parallel to the *y*-axis, we call it the vertical component.

Depending on the situation however, we need to decide which component is vertical or horizontal.

That is, *what matters is the interpretation of each component.*

Fig. 1

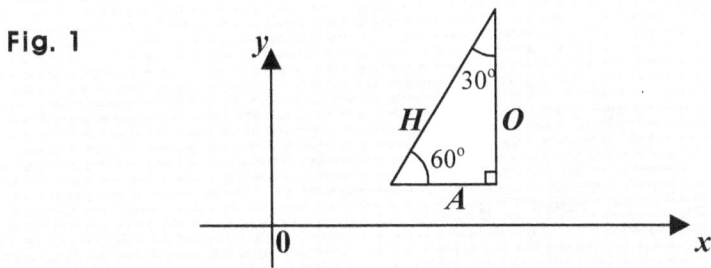

Suppose for instance, in the figure above, **O** is parallel a wall, a force is applied to the wall in the direction of **H**, and the length of **H** indicates the amount of the force.

Then, the amount of force acting on the wall vertically is **A**, and the amount of force acting on the wall horizontally is **O**. That is, of the force **H**, the vertical component acting on the wall is **A**, and the horizontal component acting on the wall is **O**.

So assuming for instance, **H** is 10 (kg·m/sec^2) acting on the wall at 30°, and referring to the figure above, we get:

The vertical component = **H·cos 60°** $= 10 \cdot \frac{1}{2} = 5$, which is **A**.

The horizontal component = **H·sin 60°** $= 10 \cdot \frac{\sqrt{3}}{2} = 5\sqrt{3}$, which is **O**.

And checking to see if they are right, we get: $A^2 + O^2 = 25 + 25 \cdot 3 = 25 \cdot 4 = 10^2 = H^2$.

Why is though, the governing angle not 30° but 60°?

The governing angle is the angle between the hypotenuse and the ***adjacent***, which is crucial. So getting the adjacent wrong, we get a wrong ratio. And thus, we want to make sure which side is the adjacent determining the governing angle or applying trig-ratios.

Examining also, the triangle above, we can notice that **sin 30°** is equal to **cos 60°**.

In fact, if two angles are *complementary* to each other, that is, the sum of the two is 90°, the sine of one angle equals the cosine of the other.

That is, if $\theta_1 + \theta_2 = 90°$, we get: **sin** θ_1 = **cos** θ_2.

In other words, we have: **cos (90° –** θ**) = sin** θ.

And of course, we have: **sin (90° –** θ**) = cos** θ, too.

What then, about **tan (90° –** θ**)**?

It is **cot** θ. So we have: **tan (90° –** θ**) = cot** θ. And also, we have: **cot = (90° –** θ**) tan** θ.

And thus, if $\theta_1 + \theta_2 = 90°$, we get: **tan** θ_1 = **cot** θ_2.

And we call those equalities trig-identities. And we can show how we get the identities using other trig-identities as **sin (**$a \pm b$**) = sin** a **· cos** b **± cos** a **· sin** b.

And we can explain them using graphs, too.

By the way, if $\theta_1 + \theta_2 = 180°$, the two angles are said to be *supplementary* to each other.

And in that case, too, we have some trig-identities. For instance, we have:

sin (180° – θ**) = sin** θ, **cos (180° –** θ**) = –cos** θ, and **tan (180° –** θ**) = –tan** θ.

And in a separate section, we will get to the details on those identities, together with the others stated above, of course.

4. What to do with Trig—Ratios 2

Multiplying a vector by a trig-ratio, we can get a component of the vector.
So extracting each component of a vector, we multiply the vector by a trig-ratio.

That is, when finding each component of a vector, we multiply the vector by the sine, the cosine, or the tangent, and alternatively, we can multiply it by the cosecant, the secant, or the cotangent.

And a trig-ratio is a ratio, too, and is nothing but a ratio.
It is a ratio between two objects, which are two sides in a right triangle.

A ratio between two objects indicates the relative amount of one object with respect to the amount of the other. So the ratio is the amount of one object relative to the other's.
So for instance, one can be twice the other, or one can be half the other.
Thus, in short, the ratio is: the amount of one / the amount of the other.

So using the ratio, we can find the amount of one for a particular amount of the other.
And we get the amount of one multiplying by the ratio the particular amount of the other.

• So to begin with, we can put **the sine** the way below:

The sine is the ratio of the opposite to the hypotenuse, in a right triangle, of course.

So assuming **O** is the opposite, and **H** is the hypotenuse, we can say that the sine is the ratio of the opposite **O** to the hypotenuse **H**.

In short, the sine is the ratio: the opposite / the hypotenuse.

Getting thus, the opposite **O** subject to a particular hypotenuse **H**, we multiply the particular hypotenuse **H** by the *ratio* of **O** to **H**, that is, **sin θ**.

So getting the opposite that is subject to a particular hypotenuse, we multiply the particular hypotenuse by the sine.

And thus, if in a right triangle, the hypotenuse is H, the opposite is O, and the *governing angle* is θ, that is, *the angle between the hypotenuse and the adjacent is θ*, we can get the opposite O the way as follows: $O = \sin \theta \cdot H = H \sin \theta$.

So for instance, if in a right triangle, the hypotenuse is 8, the opposite is O, and the *governing angle* is $30°$, that is, *the angle between the hypotenuse and the adjacent is $30°$*, we get: $O = H \sin \theta = 8 \sin 30° = 8 \cdot (1/2) = 4$.

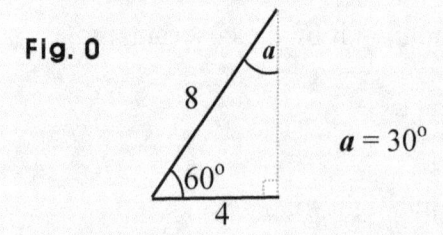

Fig. 0

$a = 30°$

And the same is true, too, for the cosine, the tangent, and all the other trig-ratios: the cosecant, the secant, and the cotangent. That is to say that we can apply the same principle applied to the sine to each of all the other trig-ratios.

• Thus, next, moving on to **the cosine**, we can put it the way below:

The cosine is the ratio of the adjacent to the hypotenuse, in a right triangle, of course.

So assuming A is the adjacent, and H is the hypotenuse, we can say that the cosine is the ratio of the adjacent A to the hypotenuse H.

In short, the cosine is the ratio: the adjacent / the hypotenuse.

Getting thus, the adjacent A subject to a particular hypotenuse H, we multiply the particular hypotenuse H by the *ratio* of A to H, that is, **cos θ**.

So getting the adjacent subject to a particular hypotenuse, we multiply the particular hypotenuse by the cosine.

And thus, if in a right triangle, the hypotenuse is H, the adjacent is A, and the *governing angle* is θ, that is, *the angle between the hypotenuse and the adjacent is θ*, we can get the adjacent A the way as follows: $A = \cos\theta \cdot H = H\cos\theta$.

So for instance, if in a right triangle, the hypotenuse is 8, the adjacent is A, and the *governing angle* is 60°, that is, *the angle between the hypotenuse and the adjacent is 60°*, we get: $A = H\cos\theta = 8\cos 60° = 8 \cdot (1/2) = 4$.

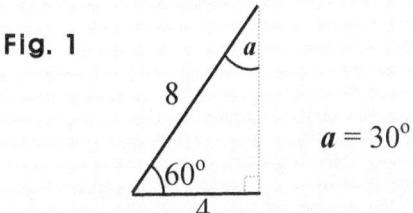

Fig. 1

$a = 30°$

• And next, moving on to **the tangent**, we can put it the way below:

The tangent is the ratio of the opposite to the adjacent, in a right triangle, of course.

So assuming O is the opposite, and A is the adjacent, we can say that the tangent is the ratio of the opposite O to the adjacent A.

In short, the tangent is the ratio: the opposite / the adjacent.

Getting thus, the opposite O subject to a particular adjacent A, we multiply the particular adjacent A by the *ratio* of O to A, that is, **tan θ**.

So getting the opposite subject to a particular adjacent, we multiply the particular adjacent by the tangent.

And thus, if in a right triangle, the opposite is O, the adjacent is A, and the *governing angle* is θ, that is, *the angle between the hypotenuse and the adjacent is θ*, we can get the opposite O the way as follows: $O = \tan\theta \cdot A = A\tan\theta$.

So for instance, if in a right triangle, the adjacent is 4, the opposite is O, and the *governing angle* is 60°, that is, *the angle between the hypotenuse and the adjacent is 60°*, we get: $O = A \tan \theta = 4 \tan 60° = 4 \cdot \sqrt{3} = 4\sqrt{3}$.

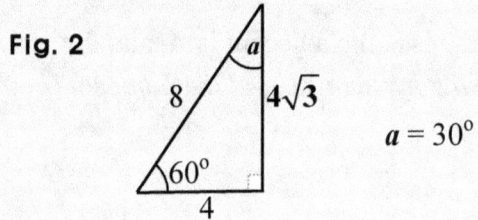

Fig. 2

$a = 30°$

- What then, about **the cosecant**?

The cosecant is the reciprocal of the sine, which is the ratio of the opposite to the hypotenuse. And thus, the cosecant is the ratio of the hypotenuse to the opposite.

So assuming O is the opposite, and H is the hypotenuse, we can say that the cosecant is the ratio of the hypotenuse H to the opposite O, and is denoted by **csc θ**.

In short, the cosecant is the ratio: the hypotenuse / the opposite.

Getting thus, the hypotenuse subject to a particular opposite, we multiply the particular opposite by the cosecant.

And thus, if in a right triangle, the hypotenuse is H, the opposite is O, and the *governing angle* is θ, that is, *the angle between the hypotenuse and the adjacent* is θ, we can get the hypotenuse H the way as follows: $H = \csc \theta \cdot O = O \csc \theta$.

So for instance, if in a right triangle, the opposite is 4, the hypotenuse is H, and the *governing angle* is 30°, that is, *the angle between the hypotenuse and the adjacent* is 30°, we get: $H = O \csc \theta = 4 \csc 30° = 4 \cdot 2 = 8$.   (sin 30° = 1/2, and **csc** 30° = 1/sin 30°.)

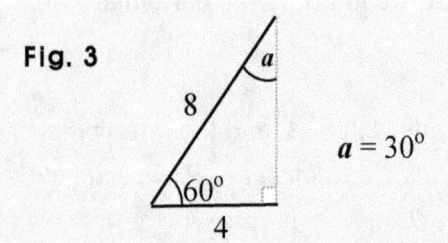

Fig. 3

$a = 30°$

And the same principle applies to the other trig-ratios: the secant and the cotangent, too.

So putting threads together, we can say that once we've set the hypotenuse and the governing angle, we actually made a right triangle.

That's because the two legs, that is, the adjacent and the opposite are subject to the hypotenuse. And vice versa. So in fact, all the three sides are subject to each other.

And more specifically:

The opposite is subject to the hypotenuse by the ratio called the sine.
And also, the opposite is subject to the adjacent by the ratio called the tangent.

The adjacent is subject to the hypotenuse by the ratio called the cosine.
And also, the adjacent is subject to the opposite by the ratio called the cotangent.

The hypotenuse is subject to the opposite by the ratio called the cosecant.
And also, the hypotenuse is subject to the adjacent by the ratio called the secant.

So in sum, assuming θ is the governing angle, H is the hypotenuse, A is the adjacent, and O is the opposite, we get:

$$O = H \cdot \sin \theta, \quad O = A \cdot \tan \theta,$$

$$A = H \cdot \cos \theta, \quad A = O \cdot \cot \theta,$$

$$H = O \cdot \csc \theta, \quad H = A \cdot \sec \theta. \qquad \text{(No one needs to memorize them though.)}$$

And thus, multiplying one of the sides by a trig-ratio appropriate, we can get the side we want. So for instance, multiplying the hypotenuse by the sine, we can get the opposite.

So assuming a vector is the hypotenuse in a right triangle, and θ is the angle the vector makes with a horizontal line as the x-axis, we can get the horizontal component multiplying the vector by $\cos \theta$.

That's because in that case, the horizontal component is in fact, the adjacent.

And thus, assuming **H** is the vector, **A** is the horizontal component, and **θ** is the angle between **H** and **A**, we get: $A = H \cdot \cos \theta$.

And also, assuming **O** is the vertical component, since **(90° – θ)** is the angle between **H** and **O**, we get: $O = H \cdot \sin \theta$.

It's simply because **θ** is the governing angle, since the angle between **H** and **A** is **θ**, because the sum of all the three angles in a triangle is 180°, **(90° – θ)** is the angle between **H** and **O**, and the angle between **A** and **O** is 90°.

So extracting the vertical component from a vector, we multiply the vector by the sine. And extracting the horizontal component, we multiply the vector by the cosine.

What is a vector though?

A vector is an object that has a magnitude and a direction. And we can specify a direction by an angle. So in short, a vector is an amount with an angle. And thus, indicating a vector, we can use an arrow with an angle. So putting for instance, an arrow of length 4 with an angle of 30° against the **x**-axis, we can put it the way below:

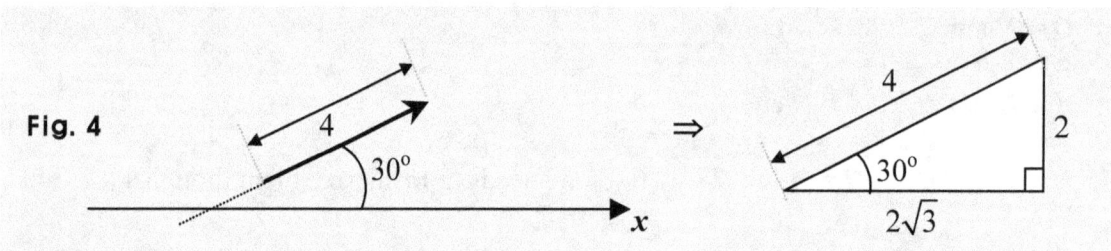

Fig. 4

So in the case above, finding the vertical component, we apply the sine of 30° to the arrow, that is, we multiply 4 by **sin 30°**. Then, we get: **4·sin 30° = 4·(1/2) = 2**.

And thus, the sine says the relative amount of the opposite with respect to the amount of the hypotenuse. What then, about the cosine?

We know that the cosine is the ratio of the adjacent to the hypotenuse. So the cosine says the relative amount of the adjacent with respect to the amount of the hypotenuse. And thus, multiplying the hypotenuse by the cosine, we get the adjacent.

And in fact, assuming the hypotenuse is 1, we can see that the cosine itself is the adjacent. It's because the cosine is the adjacent over the hypotenuse, and thus, is the adjacent, since the hypotenuse is 1.

Fig. 5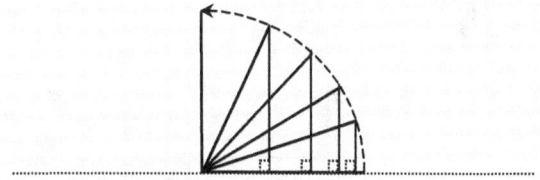

Suppose now, that in all the right triangles above, every hypotenuse is equal, and is 1, and that in each right triangle, the horizontal line segment is the adjacent.

Then, in each right triangle, the cosine itself is the adjacent. And as the governing angle increases from 0° to 90°, the adjacent decreases from 1 to 0, and so does the cosine.

And also, in each right triangle, assuming again, the hypotenuse is 1, we can see that the sine itself is the opposite. It's because the sine is the opposite over the hypotenuse, and thus, is the opposite, since the hypotenuse is 1.

So multiplying the hypotenuse by the sine, what do we get?

We get the opposite, of course.

Suppose again, that in all the right triangles above, every hypotenuse is equal, and is 1, and that in each right triangle, the horizontal line segment is the adjacent.

Then, in each right triangle, the sine itself is the opposite.
And as the governing angle increases from 0° to 90°, the opposite increases from 0 to 1, and so does the sine.

And also, we want to note that depending on the situation where the vector is applied, we need to decide which component is vertical or horizontal. That is, what matters is the interpretation of each component.

Suppose for instance, in the figure below, O is parallel to a wall, a force is applied to the wall in the direction of H, and the length of H indicates the amount of the force.

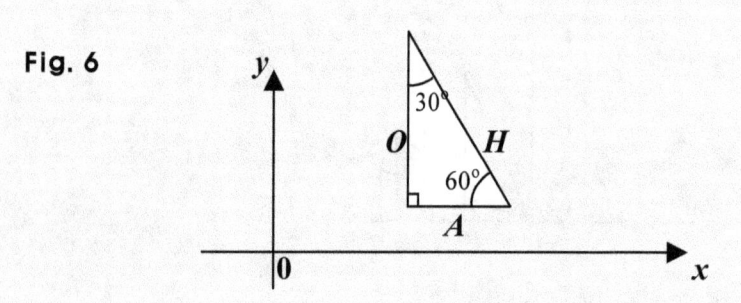

Fig. 6

Then, the amount of force acting on the wall vertically is A, and the amount of force acting on the wall horizontally is O. That is, of the force H, the vertical component acting on the wall is A, and the horizontal component acting on the wall is O.

So assuming for instance, H is 10 (kg·m/sec^2) acting on the wall at 30°, and referring to the figure above, we get:

The vertical component acting on the wall $= H\cdot\cos 60° = 10\cdot\frac{1}{2} = 5$, which is A.

The horizontal component acting on the wall $= H\cdot\sin 60° = 10\cdot\frac{\sqrt{3}}{2} = 5\sqrt{3}$, which is O.

What then, about the vector F as follows?

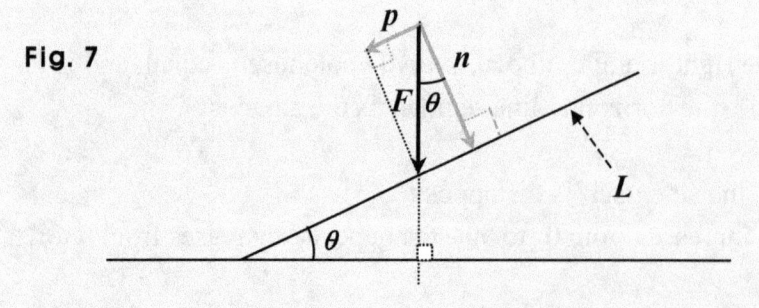

Fig. 7

Then, the force acting on the line L can be decomposed into two components.

And the two components are perpendicular to each other.

In the figure above, one is *p*, which is a force parallel to the line *L*, and the other is *n*, which is a force perpendicular to the line *L*.

Then, multiplying the vector *F* by trig-ratios as **sin** or **cos**, we can extract (find) those two components *p* and *n* from the vector *F*.

So assuming θ is the governing angle, we can get *p* and *n* the way as follows:

$p = F \sin \theta$, and $n = F \cos \theta$.

So extracting the components of a vector, we apply the sine or cosine to the vector. That is, finding each component, we multiply the vector by the sine or the cosine.

And note that the object we can apply trig-ratios to does not have to be a vector.

That is, the object we multiply by such a ratio does not have to be a vector. Such an object can be in fact, just about anything if we can put it in terms of its components perpendicular to each other. So it can be a line segment, a disk, a plane, etc.

Besides, getting each component, we often say that we get each projection, which is like a shadow. So a projection is a component of an object with an angle. For instance, we can get a projection of a circular disk the way as follows:

Fig. 8

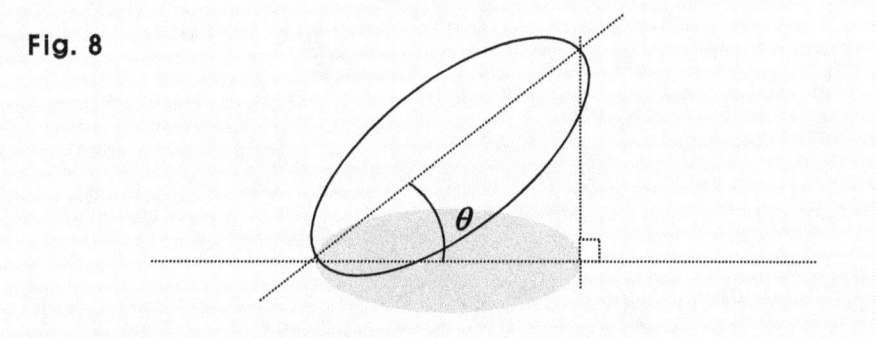

Then, the elliptic disk in gray is the projection.

And assuming **C** is the area of the circular disk, **E** is the area of the projection, that is, the area of the elliptic disk, and finding **E**, we can get it the way as follows:

$$E = C \cos \theta.$$

And also, we can get another projection of the circular disk the way as follows, too:

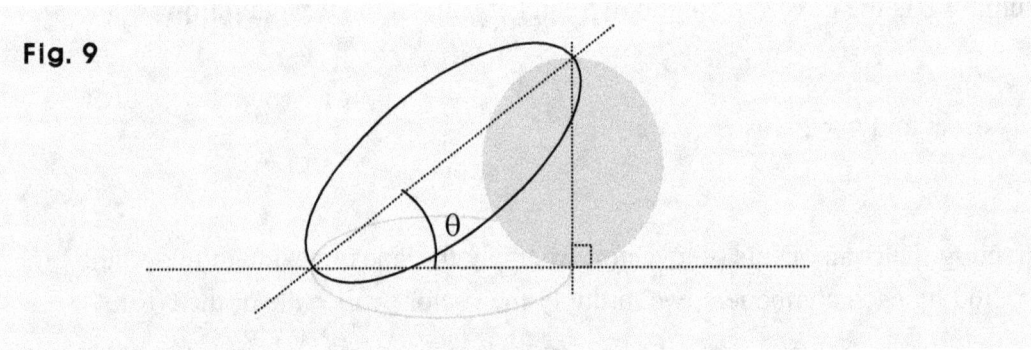

Fig. 9

Then, the other projection is an elliptic disk, too, and the two projections are perpendicular to each other.

And assuming for instance, **M** is the area of the other projection above, that is, the area of the other elliptic disk, and finding **M**, we can get it the way as follows:

$$M = C \sin \theta.$$

5. The Sine Rule

The sine is a trig-ratio, and the sine of an angle is the ratio of the opposite to the hypotenuse in a right triangle, where the hypotenuse and the adjacent makes the angle. So in short, the sine is the opposite over the hypotenuse in a right triangle. What then, is the sine rule?

It is often called the law of sines, too, and we can call it the sine formula, also. And as stated above, we get the trig-ratio called the sine from a right triangle. Using the sine rule though, we can apply it to any triangle, and thus, not a right triangle only. And by means of the sine rule, we can relate the three sides in a triangle T, for instance, in terms of the sines of all the three angles. That's not it though.

The sine rule is an expression, which is an equality, where three ratios are equal to each other, and shows that the three ratios equal a particular amount. And the particular amount is twice the radius of a circle passing through all the vertices of the triangle T. And such a circle is called a *circumcircle*. So the particular amount is the diameter of the circumcircle passing through all the vertices of the triangle T.

And each of the three ratios is a ratio of a side of the triangle T to the sine of the angle facing the side.

And thus, assuming the three angles are A, B, and C, the three sides are a, b, and c, the radius is R, and indicating the sine rule, we can put it the way below:

$$\frac{a}{\sin A} = \frac{b}{\sin B} = \frac{c}{\sin C} = 2R.$$

And since $2R$ is the diameter, assuming the diameter is D, we can put it this way too:

$$\frac{a}{\sin A} = \frac{b}{\sin B} = \frac{c}{\sin C} = D.$$

And showing the triangle T and the circumcircle, we can put them both the way below:

Fig. 0

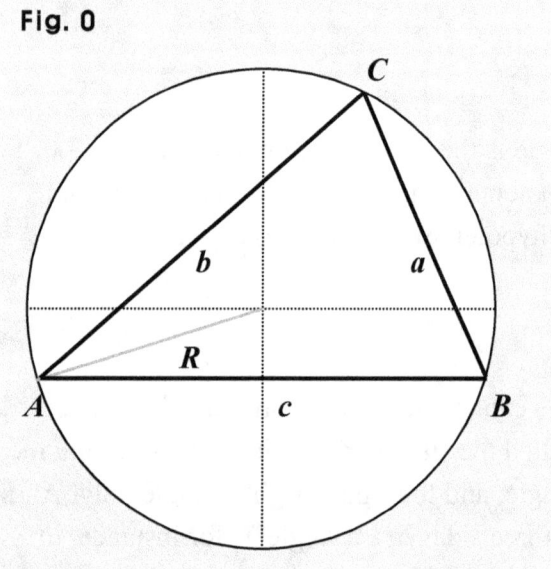

And showing briefly the proof, we can put it the way below:

$$b \sin A = a \sin B \Rightarrow \frac{a}{\sin A} = \frac{b}{\sin B}, \text{ and } c \sin A = a \sin C \Rightarrow \frac{a}{\sin A} = \frac{c}{\sin C}.$$

(And of course, we can get this, too: $b \sin C = c \sin B \Rightarrow \dfrac{b}{\sin B} = \dfrac{c}{\sin C}$.)

So we get: $\dfrac{a}{\sin A} = \dfrac{b}{\sin B} = \dfrac{c}{\sin C}.$

However, the equality above does not show the diameter, that is, $2R$.
Besides, the proof above does not cover all possible cases where the triangle T can be shaped. It covers triangles acute only, where all the three angles are between 0 and $90°$. The sine rule covers in fact, any triangle, which can be therefore, an acute triangle, an obtuse triangle where one angle is between $90°$ and $180°$, and even a right triangle, too.

So we may want to see the full proof, which covers triangles of all kinds, of course.

And more importantly, going through the full proof, we get to cover many tools we can use solving problems not only in trigonometry but other areas, too. And in fact, what's really important is that going through the full proof taking each and every step it takes, you can grow your math skills. So let's now go through the full proof.

Now, beginning with the definition of the sine rule, we can put it the way below:

Assuming A, B, and C are the three angles in a triangle inscribed in a circle of radius R, a is the side facing the angle A, b is the side facing the angle B, and c is the side facing the angle C, we get:

$\dfrac{a}{\sin A} = \dfrac{b}{\sin B} = \dfrac{c}{\sin C} = 2R$, which is called the sine rule or the sine formula.

And next, assuming T is the triangle, U is the circumcircle, and putting both in the x-y plane, we can put them the way below:

Fig. 1

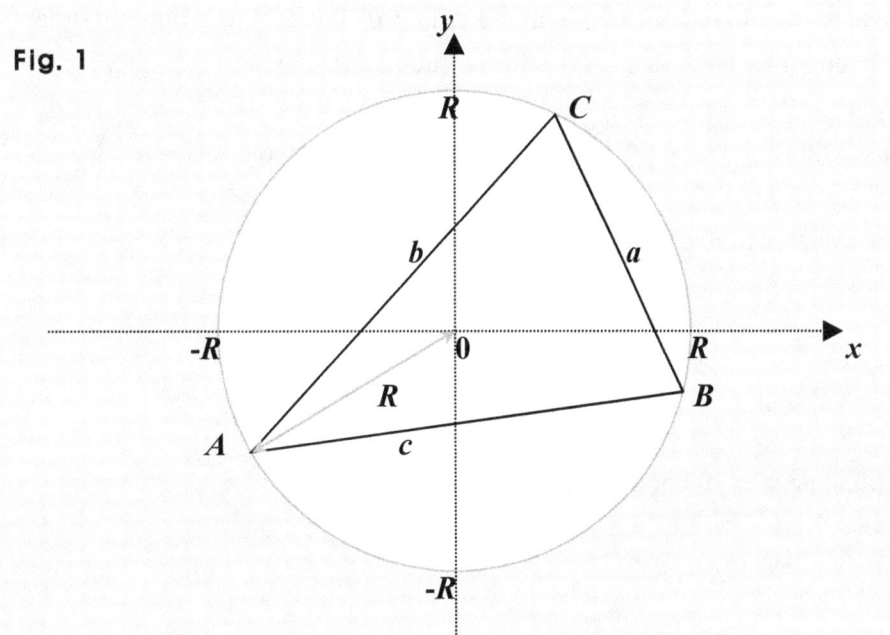

Quite often though, it is simply put this way, too: $\dfrac{a}{\sin A} = \dfrac{b}{\sin B} = \dfrac{c}{\sin C}$, which is just less the diameter, $2R$.

And in the formula above, note that one of the three angles can be obtuse as well as acute or $90°$. That is, we can have either of $0 < A < \pi$, $0 < B < \pi$, and $0 < C < \pi$.

And we are now going to see the proof, that is, how it is the case.

To begin with, no matter what triangle it may be, it can be inscribed in a circle.

In other words, every triangle has its circumscribed circle, called a circumcircle, too, which is a circle passing through all the vertices of the triangle.

That's because three line segments can make one triangle only. Four line segments though, can make more than one quadrangle, and in fact, can make infinitely many quadrangles. And the same is true for the other polygons, too. So not every polygon has a circumcircle.

And next, we want the proof to cover all triangles. We can do so producing each proof in each of three different cases where the triangle T can be.

One is a case where the angle C is an acute angle, that is: $0 < C < \pi/2$, and the other two angles are acute, too. That is, each of all the three angles is between 0 and $\pi/2$.

Another is a case where the angle C is an obtuse angle, that is: $\pi/2 < C < \pi$.

And the other is a case where the angle C is a right angle, that is: $C = \pi/2$.

• So let's now begin with the case where $0 < C < \pi/2$, that is, $0 < C < 90°$.

Then, to begin with, basically, what are we working on now?

It is a trig-ratio, and in this case, is the sine. Where then, is the trig-ratio from?

It is from a right triangle. So it's a good idea to come up with a right triangle.
And we can form one using a property of a circle. And the property is a fact that:

Fact 1: Forming a triangle using one point in a circle and two endpoints in a diameter of the circle, we get a right triangle, where the vertex angle at the one point is a right angle.

Fig. 2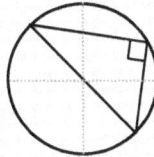

(And we are going to see the proof shortly after this full proof.)

So using **Fact 1**, we can form a right triangle, and should be able to get the sine of each angle in the triangle **T**. Where then, do we put a diameter in the circumcircle **U**?

We may want to put it so that we can use one of the three sides of the triangle **T** forming a right triangle. And also, we want to keep in mind that we want to come up with the sine of each of all the three angles in the triangle **T**.
How then, can we actually come up with the sine of each angle in the triangle **T**?

We have another fact, which is in fact, the general case to which the **fact 1** described above is a particular case. And the other fact is as follows:

Fact 2: Suppose *X*, *P* and *Q* are three points in a circle, and *X* is outside the arc *PQ*.

Fig. 3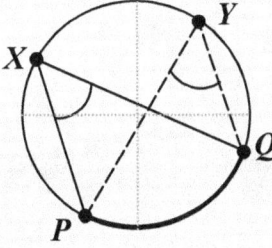

Fact 2

Then, no matter what point the point *X* may be, the angle *PXQ* is the same. So if for instance, *Y* is another point outside the arc, the angle *PYQ* equals the angle *PXQ*.

(And we will get to see the proof right after the end of this full proof.)

So keeping in mind **Fact 1** and **Fact 2** above, we may want to get back to now, the triangle **T** and the circumcircle **U**, and try putting a diameter in it.

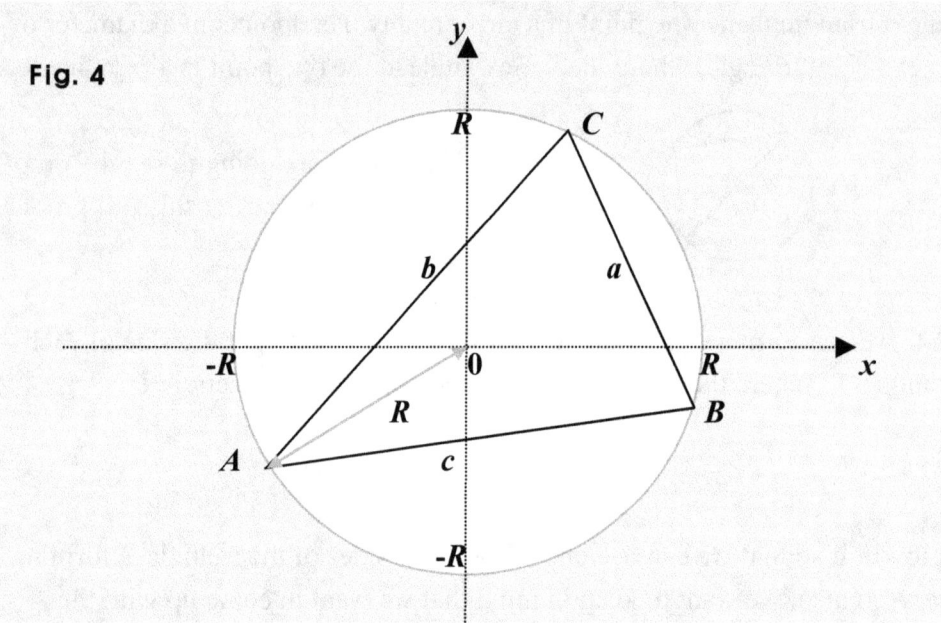

Fig. 4

Suppose first, we use the side *c* to come up with one of the three ratios in the sine rule. What ratio then, should we come up with?

It will be the ratio of the side *c* to the sine of the angle *C*, that is: $\dfrac{c}{\sin C}$.

So we may want to put a diameter in the circle so that one of its endpoints can be a vertex in the triangle *T*. And using the vertex *A* as the endpoint, we can get:

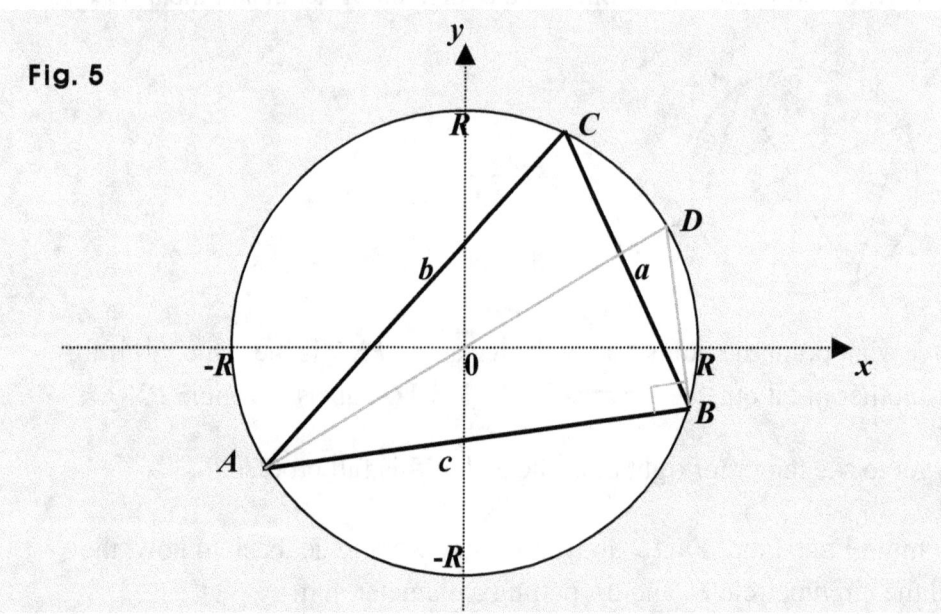

Fig. 5

And next, assuming the line segment $\overline{AD}$ passes through the center, we get: $\overline{AD} = 2R$.

So taking the sine of the angle **D**, that is, taking **sin D**, we get: $\sin D = \frac{c}{2R}$.

And next, using **Fact 2** stated above, we can say that the two angles **C** and **D** are the same.

So we get: **sin C = sin D**. And thus, we get: $\sin C = \dfrac{c}{2R} \Rightarrow 2R = \dfrac{c}{\sin C}$.

And due to **Fact 2** above, we can get the same using **B** as the endpoint, too:

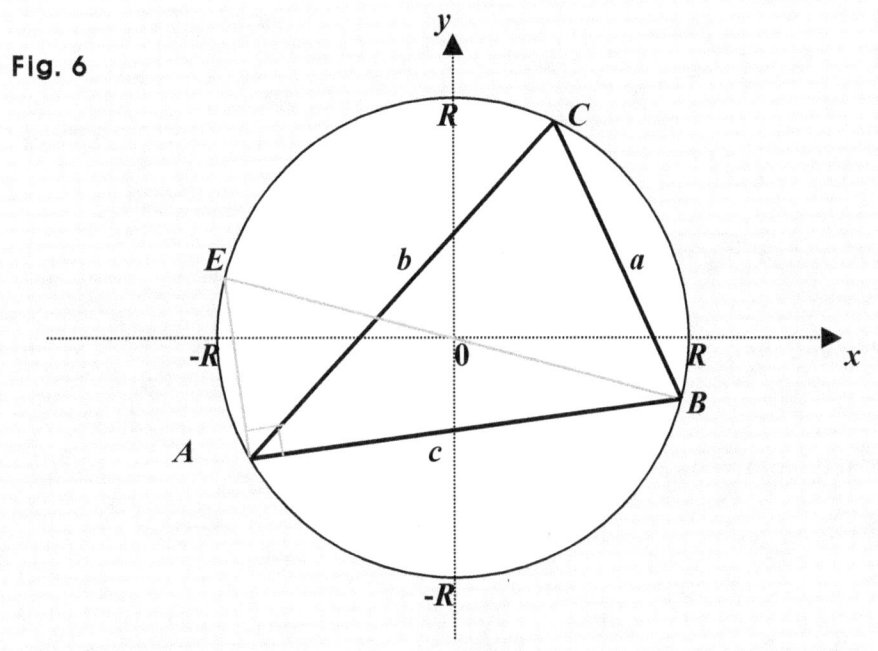

Fig. 6

Then first, assuming the line segment $\overline{EB}$ passes through the center, we get: $\overline{EB} = 2R$.

And thus, taking **sin E**, we get: $\sin E = \frac{c}{2R}$.

And next, using the **fact 2** above, we can say that the two angles **C** and **E** are the same.

So we get: **sin C = sin E**.

And next, using each of **C** and **B** as an endpoint of a diameter in the circle, we can get:

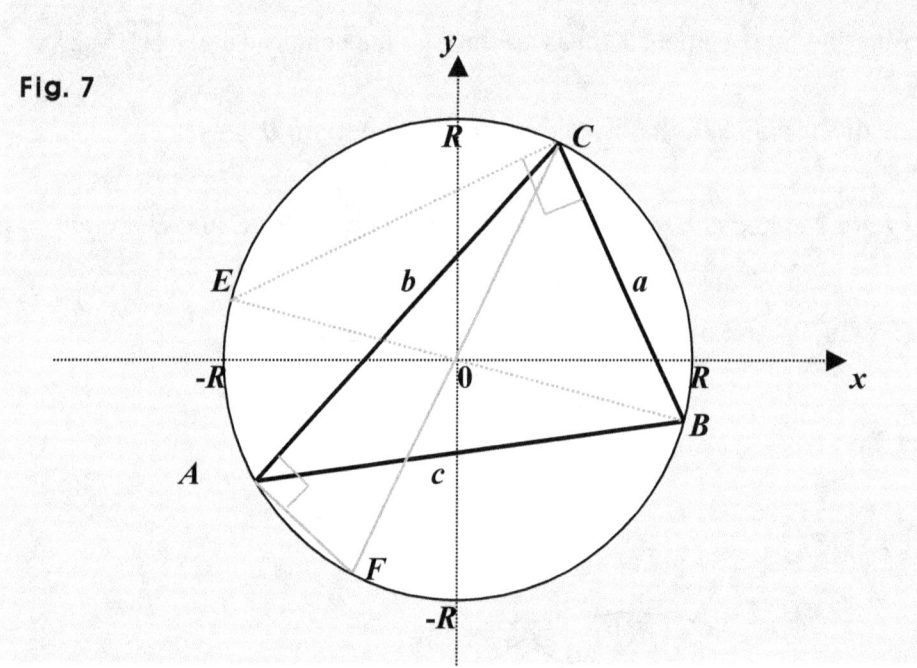

Fig. 7

So next:

Assuming $\overline{CF} = 2R$, and taking **sin F**, we get: $\sin F = \frac{b}{2R}$.

Assuming $\overline{EB} = 2R$, and taking **sin E**, we get: $\sin E = \frac{a}{2R}$.

And next, due to **Fact 2** above:

Two angles **B** and **F** are equal, so we get: $\sin B = \sin F$.

Two angles **A** and **E** are equal, so we get: $\sin A = \sin E$.

So we get: $\sin B = \dfrac{b}{2R} \Rightarrow 2R = \dfrac{b}{\sin B}$, and $\sin A = \dfrac{a}{2R} \Rightarrow 2R = \dfrac{a}{\sin A}$.

And thus, putting threads together, we get: $\dfrac{a}{\sin A} = \dfrac{b}{\sin B} = \dfrac{c}{\sin C} = 2R$.

• And let's next, move on to the case where $\pi/2 < C < \pi$, that is, **$90^\circ < C < 180^\circ$**.

So one angle in a triangle is an obtuse angle. That is, the triangle **T** is an obtuse triangle. So putting in the **x-y** plane, the obtuse triangle **T**, along with its circumcircle **U**, we get:

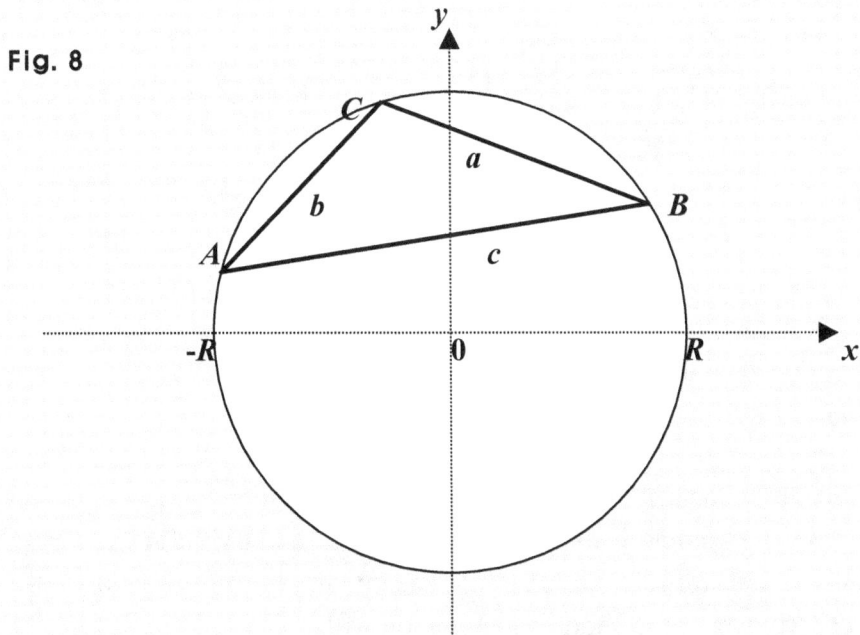

Fig. 8

As in the case above, we are basically, working on a trig-ratio, which in this case, is the sine, which is from a right triangle. So it's a good idea to come up with a right triangle.

And we can form one using a property in a circle, which is a fact, and the fact says:

Fact 3: Forming a triangle using one point in a circle and two endpoints in a diameter of the circle, we get a right triangle, where the vertex angle at the one point is a right angle.

And thus, using **Fact 3**, we can form a right triangle, and should be able to get the trig-ratios, that is, the sines. Where then, do we put a diameter in the circumcircle U?

We may want to put it so that we can use one of the sides in the triangle T forming a right triangle. And also, we want to keep in mind that we want to come up with the sines of all the three angles in the triangle T.

And we can actually come up with each sine using two more facts, one of which says:

Fact 4: Assuming P and Q are two vertex-angles facing each other in a quadrangle that has a circumcircle, we get: $P + Q = \pi$.

Fig. 9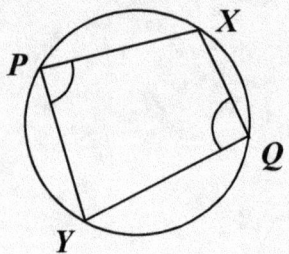

So we get: $X + Y = \pi$, too.

And in fact, we have: $X + Y + P + Q = 2\pi$.

And the other fact is in fact, a trig identity where **sin $(\pi - \theta) = \sin \theta$.**
In other words, assuming: $\theta_1 + \theta_2 = \pi$, we get: **sin $\theta_1 = \sin \theta_2$.**
So referring to the quadrangle above, we get: **sin $P = \sin Q$.**

So keeping in mind the three facts above, **Fact 3**, **Fact 4**, and the identity above, we may
want to get back to now, the triangle T and the circumcircle U, and try putting a
diameter in it. Assuming first, we use the side c to come up with one of the ratios in the
equality we want to prove, what ratio should we come up with?

It will be the ratio of the side c to the sine of the angle C, that is: $\dfrac{c}{\sin C}$.

So we may want to put a diameter in the circumcircle U so that one of its endpoints can
be a vertex in the triangle T. And using the vertex A as the endpoint, we can get:

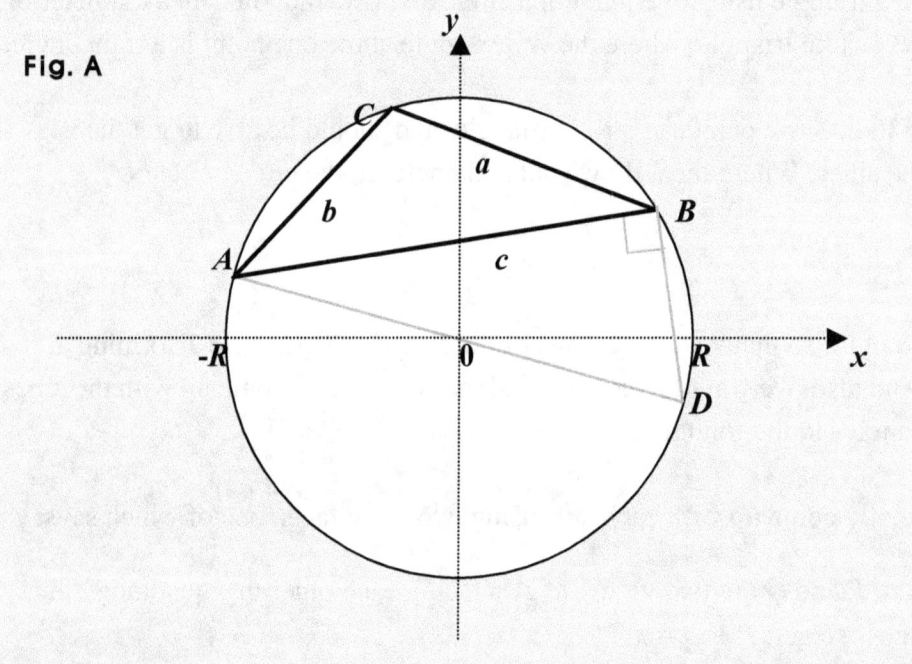

Fig. A

So next, assuming the line segment $\overline{AD}$ passes through the center, we get: $\overline{AD} = 2R$.

And thus, taking the sine of the angle D, that is, taking $\sin D$, we get: $\sin D = \frac{c}{2R}$.

And next, using the **fact 4** stated above, we get: $C + D = \pi$.

So next, using the trig-identity above, we get: $\sin C = \sin D$.

And thus, we get: $\sin C = \dfrac{c}{2R} \Rightarrow 2R = \dfrac{c}{\sin C}$.

And due to **Fact 3** above, we can get the same using B as the endpoint of a diameter, too:

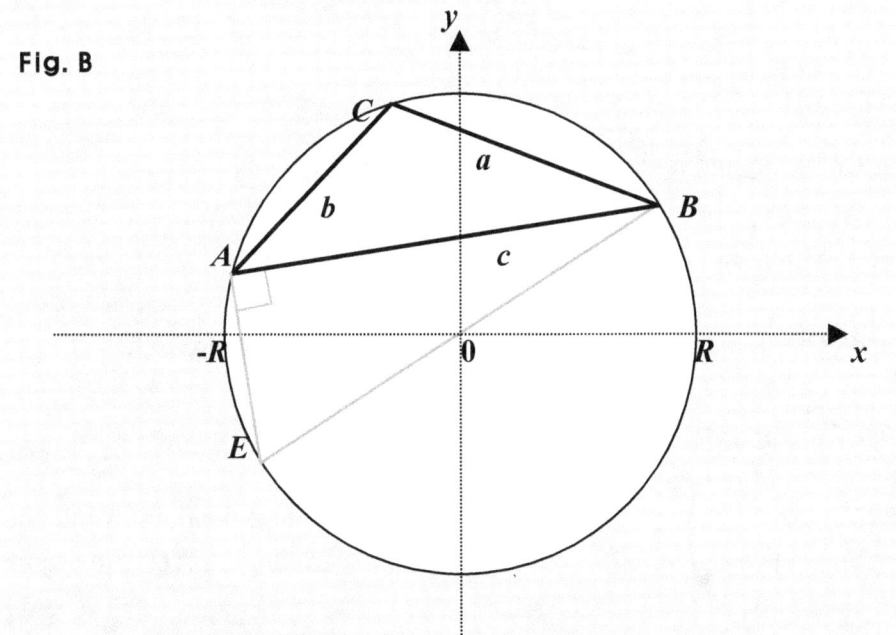

Fig. B

Then first, assuming: $\overline{EB} = 2R$, and taking $\sin E$, we get: $\sin E = \frac{c}{2R}$.

And next, using the **fact 4**, we get: $C + E = \pi$.

So next, using: $\sin (\pi - \theta) = \sin \theta$, we get: $\sin C = \sin E$.

And thus, we get: $\sin C = \dfrac{c}{2R} \Rightarrow 2R = \dfrac{c}{\sin C}$.

And next, using the vertex **C** as an endpoint of a diameter in the circle **U**, we can get:

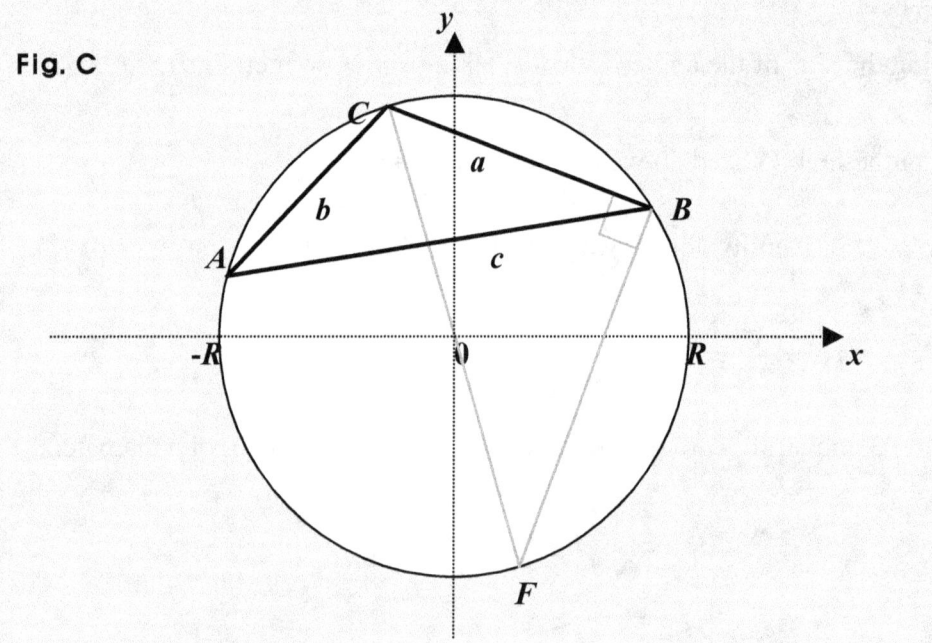

Fig. C

So next, assuming $\overline{CF} = 2R$, and taking **sin F**, we get: $\sin F = \frac{a}{2R}$.

And we have **Fact 2**, which says:

Suppose **X**, **P** and **Q** are three points in a circle, and **X** is outside the arc **PQ**.

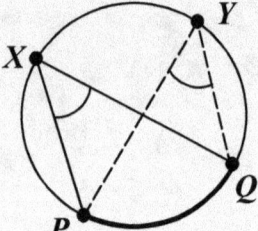

Fig. D

Then, no matter what point the point **X** may be, the angle **PXQ** is the same.

So we can say that the angle **PXQ** equals the angle **PYQ**.

And thus, getting back to **Fig. C** above, and using the fact above, which is **Fact 2**, we can say that two angles **A** and **F** are equal, so we get: **sin A = sin F**.

And thus, we get: $\sin A = \dfrac{a}{2R} \Rightarrow 2R = \dfrac{a}{\sin A}$.

And again, using the vertex C as an endpoint of a diameter in the circle U, we can get:

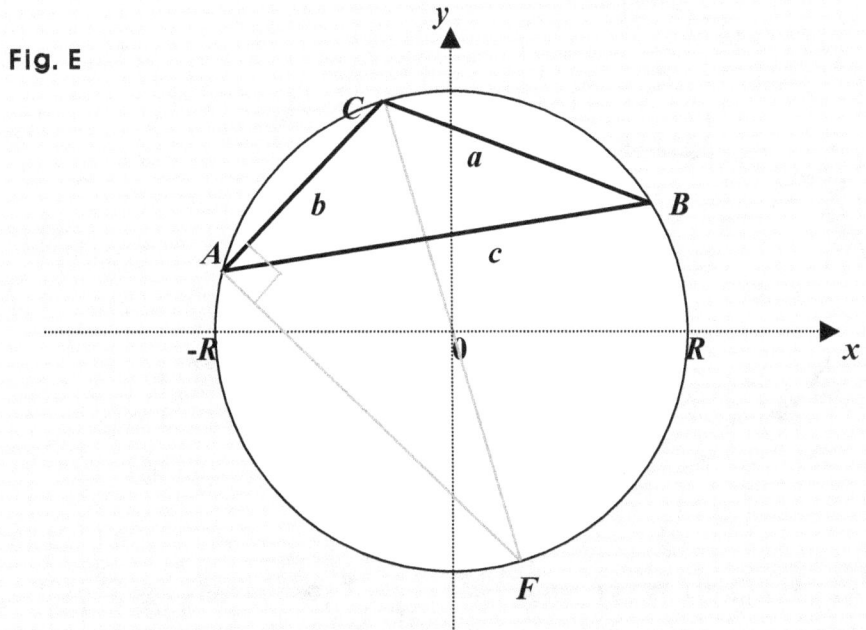

Fig. E

So next, assuming again, $\overline{CF} = 2R$, and taking $\sin F$, we get this time: $\sin F = \frac{b}{2R}$.

And thus, using **Fact 2** we have just used above, we can say that two angles B and F are equal, so we get: $\sin B = \sin F$.

So we get: $\sin B = \dfrac{b}{2R} \Rightarrow 2R = \dfrac{b}{\sin B}$.

And thus, putting threads together, we get: $\dfrac{a}{\sin A} = \dfrac{b}{\sin B} = \dfrac{c}{\sin C} = 2R$.

• And let's next, move on to the last case where $C = \pi/2$, which is $90°$.

That is, one angle in a triangle is a right angle. In other words, the triangle T is a right triangle.

So putting in the x-y plane the right triangle T, along with its circumcircle U, we can put it the way below:

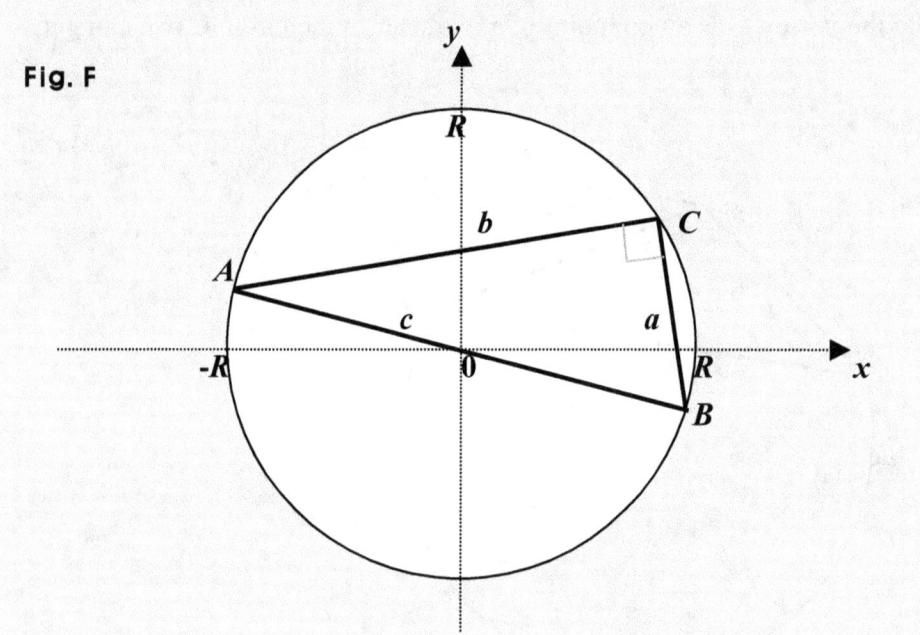

Fig. F

Then first, we know: $C = \pi/2$. So we get: $\sin C = \sin \pi/2 = 1 \Rightarrow \sin C = 1$.

And assuming the line segment $\overline{AB}$ passes through the center, we get: $\overline{AB} = 2R$.

So we get: $2R = c$. And doing some algebra on it, we can get:

$$2R = c \Rightarrow 2R = \frac{c}{1} = \frac{c}{\sin C} \quad \text{since we have: } \sin C = 1. \text{ And thus, we can get: } 2R = \frac{c}{\sin C}.$$

And next, taking simply the sines of the angles A and B, we get:

$$\sin A = \frac{a}{c} \Rightarrow c = \frac{a}{\sin A}, \text{ and } \sin B = \frac{b}{c} \Rightarrow c = \frac{b}{\sin B}. \quad \text{And we know: } c = 2R.$$

So we get: $\dfrac{a}{\sin A} = \dfrac{b}{\sin B} = \dfrac{c}{\sin C} = 2R$.

• And let's next, move on to the proofs of the facts we used covering the proof above.

To begin with, we have **Fact 1**, which is as follows:

• **Fact 1**: Suppose *X*, *P* and *Q* are three points in a circle, and *X* is outside the arc *PQ*.

Fig. G

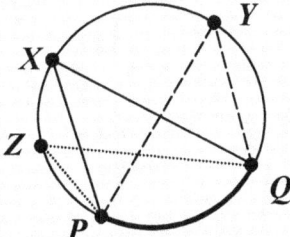

Then, no matter what point the point *X* may be, the angle *PXQ* is the same. If for instance, therefore, *Y* and *Z* are two other points outside the arc, but are in the circle, and if ∠*PZQ* indicates the angle *PZQ*, we get: ∠*PZQ* = ∠*PXQ* = ∠*PYQ*.

So now, to see how it is the case, let's take a closer look.

Fig. H

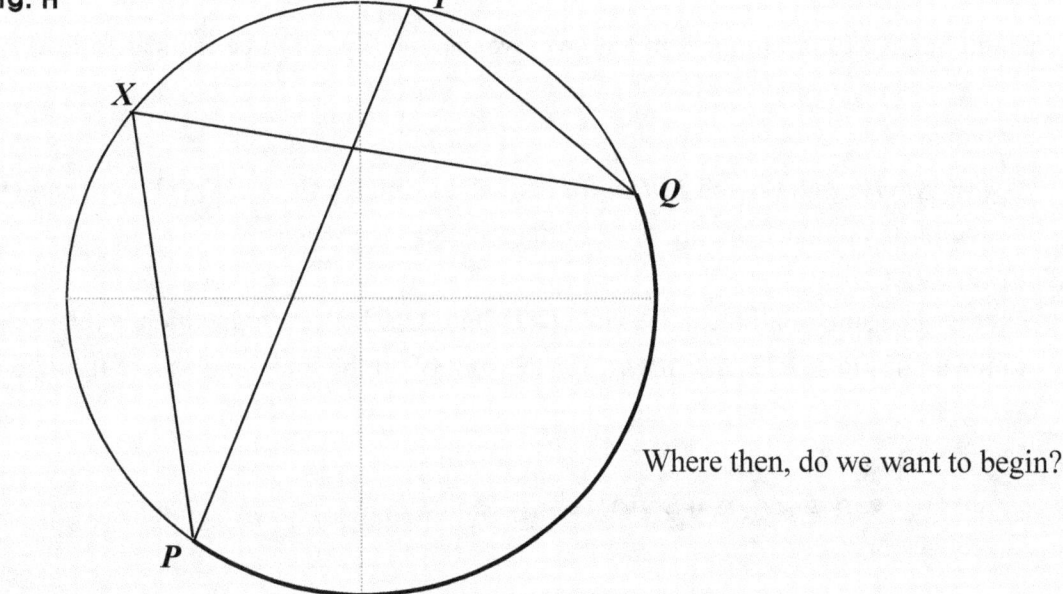

Where then, do we want to begin?

Putting some radii in it, we can see better what we can do about it.

And using symbols for words used, we can make ideas quicker or clearer to understand as well as simpler to understand. For instance, we can use Δ as the symbol for a triangle, and use ∠ as the symbol for an angle. And also, it's a good idea to add labels as necessary. So doing the suggestions above, we can get:

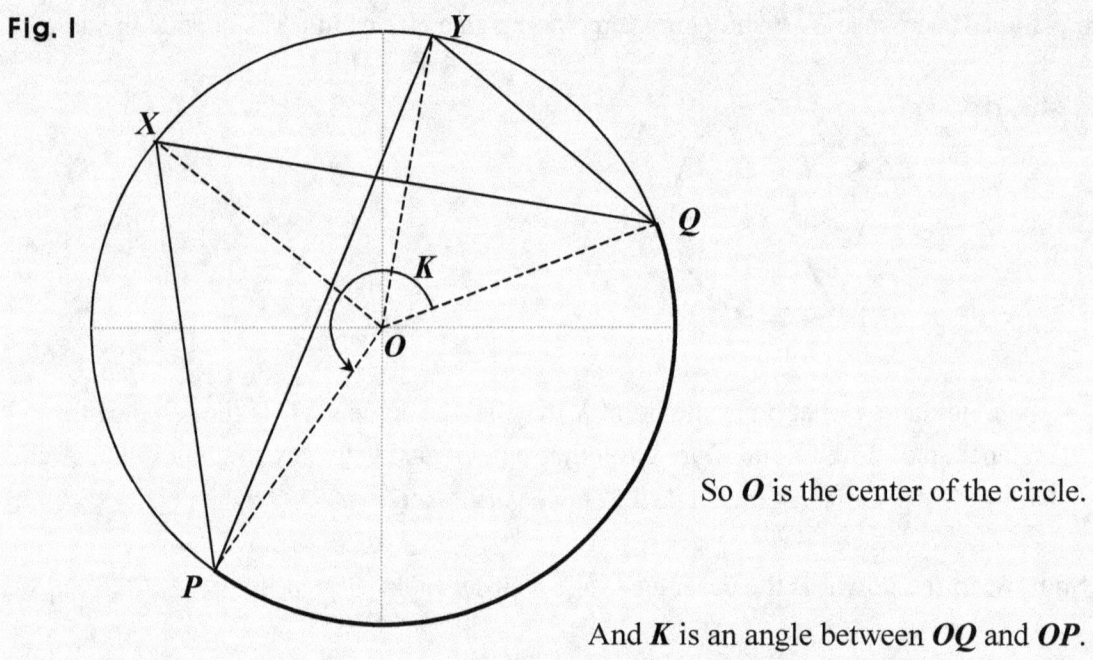

Fig. I

So **O** is the center of the circle.

And **K** is an angle between **OQ** and **OP**.

Then, to begin with, from the figure above, we can readily get three facts as follows:

Fact A:

Δ**QOY**, Δ**POY**, Δ**QOX**, and Δ**POX** are isosceles triangles since **OX**, **OY**, **OP**, and **OQ** are the radii, and thus, are the same in magnitude.

Fact B: The sum of all the angles in Δ**QOY** and Δ**POY** is: 2π rad, which is the sum of all the angles in Δ**QOX** and Δ**POX**, for the sum of all the three angles in a triangle is π.

And **Fact C**:

$\angle QOY + \angle POY = \angle QOX + \angle POX =$ the angle **K** above.

Then first, from **Fact B** and **Fact C**, we can get another fact below:

Fact D: $\angle OPY + \angle PYQ + \angle YQO = \angle OPX + \angle PXQ + \angle XQO$.

That's because $\angle OPY + \angle PYQ + \angle YQO = 2\pi - K = \angle OPY + \angle PYQ + \angle YQO$.

And next, from **Fact A**, we can get another fact below:

Fact E: $\angle YQO + \angle OPY = \angle PYQ$, and $\angle XQO + \angle OPX = \angle PXQ$.

It's because $\angle YQO = \angle QYO$, $\angle OPY = \angle OYP$, $\angle XQO = \angle QXO$, and $\angle OPX = \angle OXP$.

So next, putting **Fact E** into **Fact D**, we can get: $2\angle PYQ = 2\angle PXQ$.

And therefore, we get: $\angle PYQ = \angle PXQ$.

What if though, the point **Y** is put the way below?

Fig. J

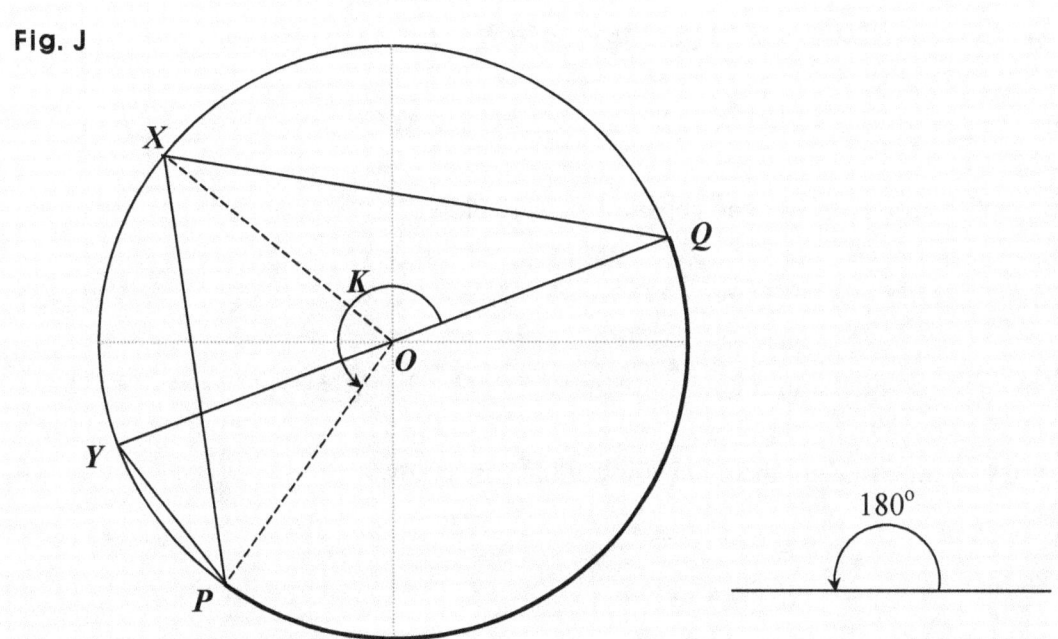

That is, **QY** is not just a line segment connecting **Q** and **Y** but a diameter, too, in the circle. Then, to begin with, from the figure above, we can get readily three facts below:

Fact F: ΔPOY, ΔQOX, and ΔPOX are isosceles triangles since **OX**, **OY**, **OP**, and **OQ** are the radii, and thus, are the same in length.

Fact G: The sum of the angle in the line segment **YQ** and all the angles in ΔPOY is: 2π rad, which is the sum of all the angles in ΔQOX and ΔPOX, because the angle in a line segment is $180°$, which is π rad, and more specifically, the angle **QOY** is $180°$.

So we get **Fact H**, which is: $\angle QOY + \angle POY = 180° + \angle POY = \angle QOX + \angle POX$.

Then first, from the fact G and the fact H, we can get another fact below:

Fact I: $\angle OPY + \angle PYQ = \angle OPX + \angle PXQ + \angle XQO$.   How come?

First, from **Fact G**, we can get: $2\pi - (\angle OPX + \angle PXQ + \angle XQO) = K$.

And also, we can get: $2\pi - (\angle OPY + \angle PYQ) = K$.

So we get: $2\pi - (\angle OPY + \angle PYQ) = 2\pi - (\angle OPX + \angle PXQ + \angle XQO)$.

Thus, we get: $\angle OPY + \angle PYQ = \angle OPX + \angle PXQ + \angle XQO$.

And next, from **Fact F**, we can get another fact below:

Fact J: $\angle OPY = \angle PYQ$, and $\angle XQO + \angle OPX = \angle PXQ$.

So next, putting **Fact J** into **Fact I**, we get: $2\angle PYQ = 2\angle PXQ \Rightarrow \angle PYQ = \angle PXQ$.

What if this time though, the point Y is put the way below?

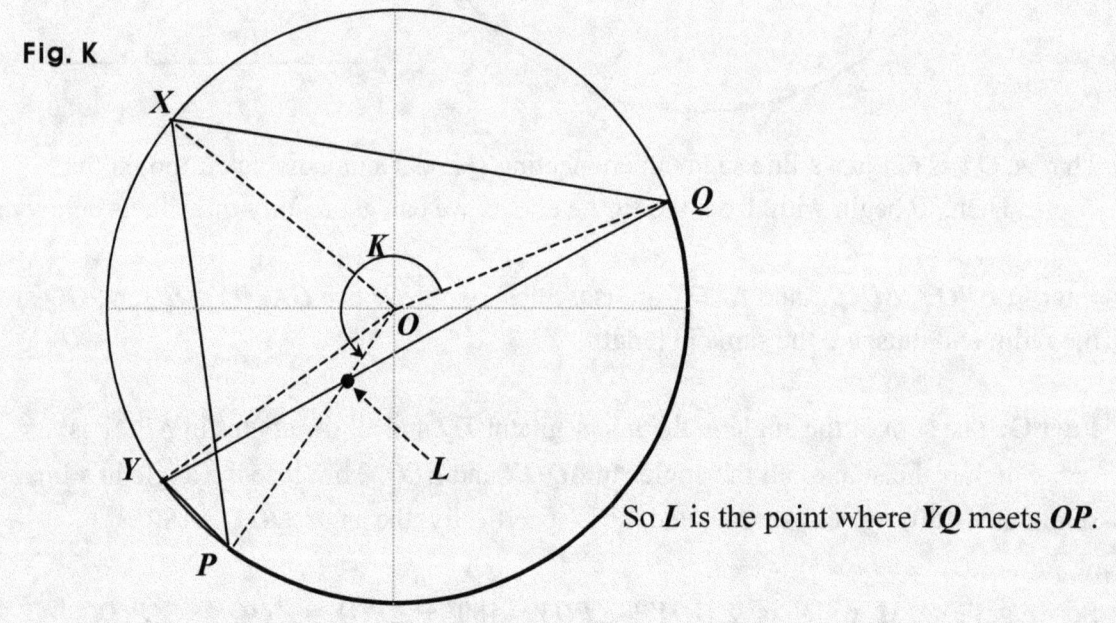

Fig. K

So L is the point where YQ meets OP.

Then, to begin with, from the figure above, we can readily get four facts as follows:

Fact K: ΔQOY, ΔPOY, ΔQOX, and ΔPOX are isosceles triangles since OX, OY, OP, and OQ are the radii, and thus, have the same length.

Fact L: The sum of all the angles in ΔQOX and ΔPOX is: 2π rad, since the sum of all the three angles in a triangle is π rad.

Fact M: We have: $\angle PYQ + \angle OPY = \angle YLO$.

And **Fact N**: We have: $\angle LOQ + \angle OQL = \angle YLO$.

Then first, from **Fact K**, we get: $\angle OYP = \angle OPY$, since ΔPOY is an isosceles triangle.

So we get: $\angle OPY = \angle PYQ + \angle OYL$.

And thus, from **Fact M**, we get:

$$\angle YLO = \angle PYQ + \angle OPY = \angle PYQ + \angle PYQ + \angle OYL = 2\angle PYQ + \angle OYL$$

$$\Rightarrow \angle YLO = 2\angle PYQ + \angle OYL.$$

So next, from **Fact N**, we get: $2\angle PYQ + \angle OYL = \angle LOQ + \angle OQL$.

And also, from **Fact K**, we get: $\angle OYL = \angle OQL$, since ΔQOY is an isosceles triangle.

So we get: $2\angle PYQ + \angle OYL = \angle LOQ + \angle OYL$. Thus, we get another fact below:

Fact O: $2\angle PYQ = \angle LOQ$.

And also, from the fact L, we can get another fact below:

Fact P: $\angle LOQ = \angle OPX + \angle PXQ + \angle XQO$.   How come though?

First, we know: the sum of all the angles in ΔQOX and ΔPOX is: 2π rad.   So we get:
$2\pi - (\angle OPX + \angle PXQ + \angle XQO) = K \Rightarrow 2\pi - K = \angle OPX + \angle PXQ + \angle XQO$.

And also, we have: $2\pi - K = \angle LOQ$.

So we get: $\angle LOQ = \angle OPX + \angle PXQ + \angle XQO$, which is **Fact P**.

And next, from **Fact K**, we get: $\angle OPX = \angle OXP$, and $\angle OQX = \angle OXQ$.

So we get: $\angle OPX + \angle PXQ + \angle XQO = 2\angle PXQ$.

And we have **Fact P**: $\angle LOQ = \angle OPX + \angle PXQ + \angle XQO$.

So we get: $\angle LOQ = 2\angle PXQ$.

And also, we have **Fact O**: $2\angle PYQ = \angle LOQ$.

So we get: $2\angle PYQ = 2\angle PXQ$. And thus, we get: $\angle PYQ = \angle PXQ$.

Besides, in the processes above, we have found an important fact we can use solving many problems. And the fact is as follows:

First, we know in the figure above, the point O is the center of the circle.
So we call the angle POQ a central angle.
More specifically, $\angle POQ$ is called the central angle for the arc PQ.

And also, we can call the angle PXQ an angle of circumference.
More specifically, $\angle PXQ$ can be called the angle of circumference for the arc PQ.

Now, the important fact is that the angle of circumference for the arc PQ is half the central angle for the arc PQ.

And we know PQ is an arbitrary arc. So we can say that:

• The central angle for an arc is twice the angle of circumference for the arc.

And we can simply prove the fact above the way below, too:

Fig. L

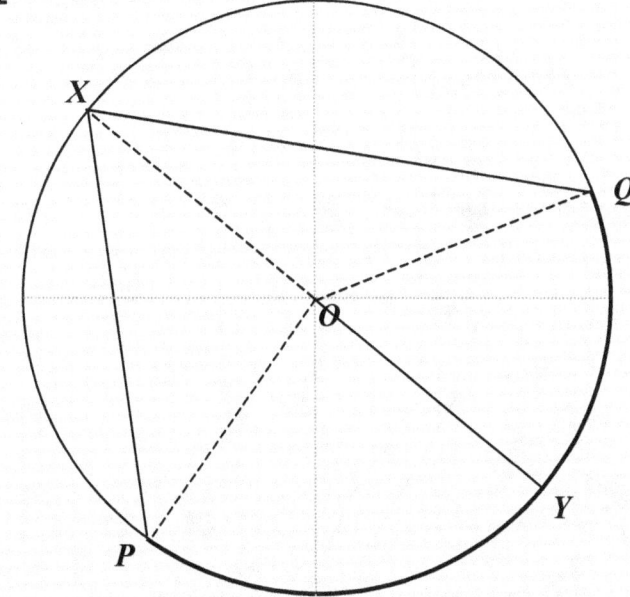

First, we know the two triangles **ΔPXO** and **ΔQXO** are isosceles triangles.

So we get: $\angle POY = \angle OPX + \angle OXP = 2\angle OXP$.

And also, we get: $\angle QOY = \angle OQX + \angle OXQ = 2\angle OXQ$.

Thus, we get: $\angle POQ = \angle POY + \angle QOY = 2\angle OXP + 2\angle OXQ = 2(\angle OXP + \angle OXQ)$.

And we have: $\angle PXQ = \angle OXP + \angle OXQ$.

So we get: $\angle POQ = 2\angle PXQ$.

And we can notice that if a circle is partitioned into many arcs, the sum of all the angles of circumference for all the arcs is 180°.

Fig. M

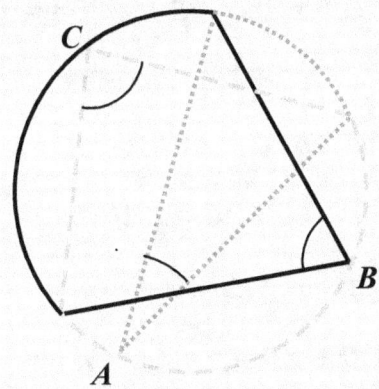

So the circle on the left is partitioned into 3 arcs.

One is solid, another is dotted, and the other is dashed.

And the sum of the three angles **A**, **B**, and **C** is 180°.

That is, we get: $A + B + C = \pi$.

And let's now move on to the next fact. And the fact says:

• The angle of circumference for a half circle is a right angle, that is, 90°.

So assuming for instance, **O** is the center of a circle, **AB** is a diameter, and **C** is an angle of circumference for a half circle, that is, the arc **AB**, we can put them all the way below:

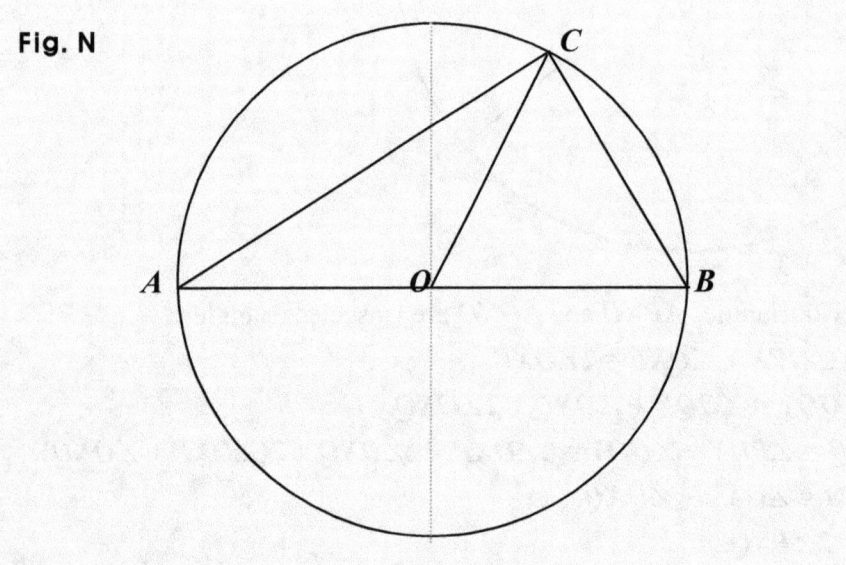

Fig. N

Then, we can begin with the two facts below:

Fact Q: △*AOC* and △*COB* are isosceles triangles since *OA*, *OC*, and *OB* are the radii.

Fact R: ∠*CAO* + ∠*ACO* = ∠*COB*, ∠*BCO* + ∠*CBO* = ∠*COA*, and ∠*COB* + ∠*COA* = π.

Then first, from the fact Q, we can get: ∠*CAO* = ∠*ACO*, and ∠*BCO* = ∠*CBO*.

So we get: ∠*CAO* + ∠*ACO* = ∠*COB* ⇒ ∠*COB* = 2∠*ACO* since ∠*CAO* = ∠*ACO*.

Also, we get: ∠*BCO* + ∠*CBO* = ∠*COA* ⇒ ∠*COA* = 2∠*BCO* since ∠*BCO* = ∠*CBO*.

So we get: ∠*COB* + ∠*COA* = 2∠*ACO* + 2∠*BCO* = 2(∠*ACO* + ∠*BCO*) = 2∠*ACB* = π.

And thus, we get: $\angle ACB = \pi/2$, which is 90°.

How come though, we get: $\angle CAO + \angle ACO = \angle COB$?

We know that angle in the line segment AB is π.
So first, we get: $\angle COA + \angle COB = \pi$.

And next, we have: $\angle CAO + \angle ACO + \angle COA = \pi$, since the sum of all the three angles in a triangle is π. So we get: $\angle CAO + \angle ACO + \angle COA = \angle COA + \angle COB$.

And thus, we get: $\angle CAO + \angle ACO = \angle COB$.

And moving on to the next fact, **Fact 4** covered earlier, we have:

Fact 4: Assuming P and Q are two vertex-angles facing each other in a quadrangle that has a circumcircle, we get: $P + Q = \pi$.

Fig. O

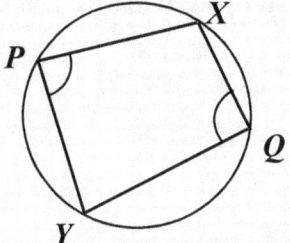

So we get: $X + Y = \pi$, too.

And in fact, we have: $X + Y + P + Q = 2\pi$.

That's because a quadrangle can be made of two triangles:

Fig. P

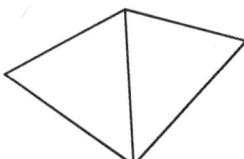 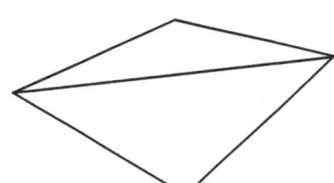

And we want to get the proof of the fact that: $P + Q = \pi$. So let's now take a closer look.

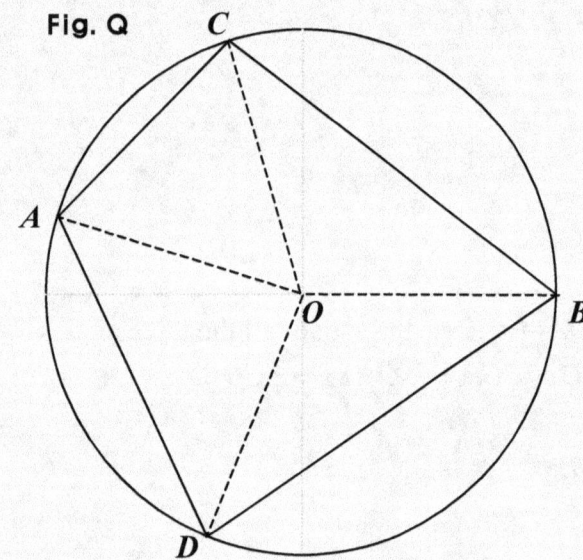

Fig. Q

To begin with, we can partition
a quadrangle into triangles.

And we are working with a circle, too.

And what determines a circle is
the radius and the center.
So we may want to try using the radii
and the center partitioning the quadrangle.

Then first, we can notice that the quadrangle is made of isosceles triangles, which are **ΔOBC, ΔOCA, ΔOAD,** and **ΔODB**.

In an isosceles triangle, at least two angles are the same. So we can use such a property. That is, labeling or specifying those angles, we can see better what we can do.

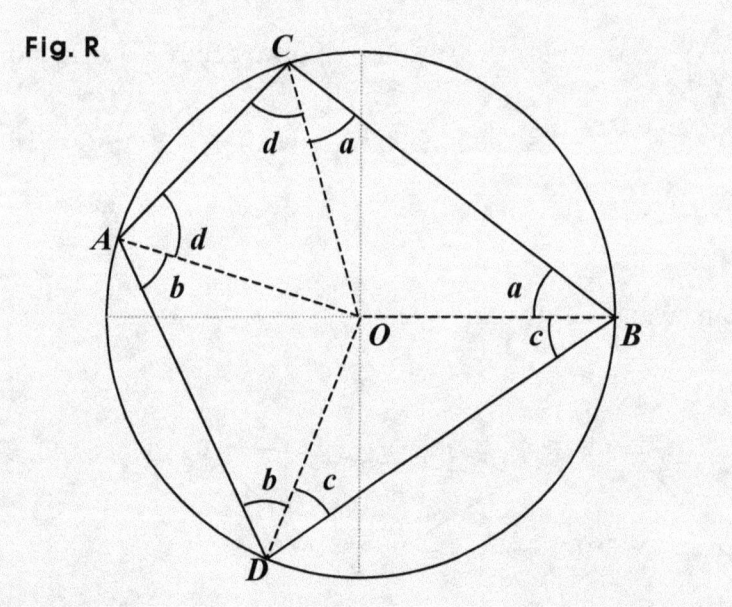

Fig. R

We can notice that the angle C is made of a and d, and that the angle D is made of b and c, which means, the two angles altogether are made of a, b, c, and d, which are all kinds of angles there.

And what we want to show is that in a quadrangle with its circumcircle, the sum of the two angles facing each other is π.

So we can reasonably expect that if we can put together all the angles around the center of the circle, that is, the point O, the sum of all the angles can be 2π.

How then, can we put them together around the center?

We can use the fact that all the radii radiate from the center, and the fact that all the equal sides in all the isosceles triangles are the radii.

And also, we have a fact that a sum of two internal angles is the same as an external angle in a triangle.

More specifically, among the three internal angles in a triangle, the sum of two internal angles is equal to the external angle supplement to the other internal angle.
Refer to the figure below:

Fig. S

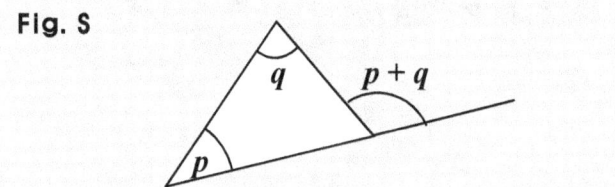

So using the fact above, and extending one of the two equal sides in each of the isosceles triangles, we can put together all the angles around the center.

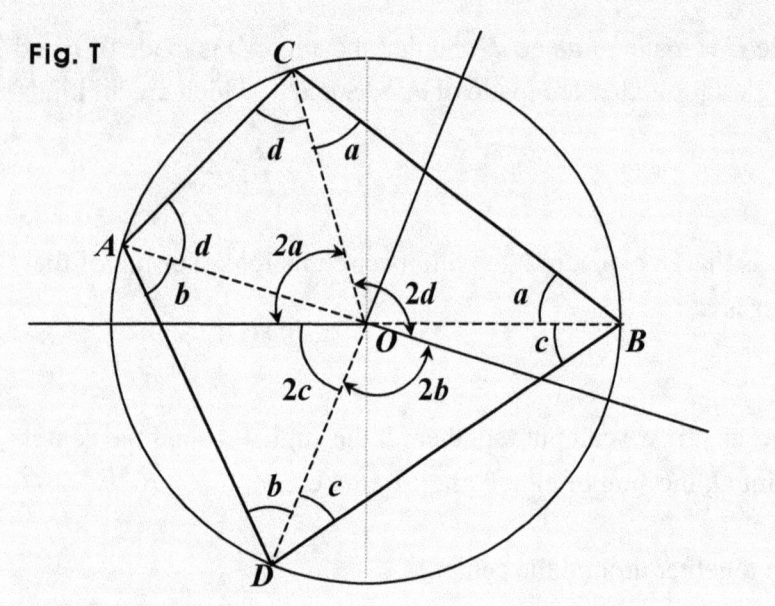

Fig. T

Thus, we get: $2a + 2b + 2c + 2d = 2(a + b + c + d) = 2\pi$.

So we get: $a + b + c + d = 2\pi$.

And thus, we get: $A + B = \pi$, and $C + D = \pi$.

• And finally, the other fact is in fact, a trig identity where $\sin(\pi - \theta) = \sin \theta$.

In other words, assuming: $\theta_1 + \theta_2 = \pi$, we get: $\sin \theta_1 = \sin \theta_2$.

So referring to the quadrangle above, we get: $\sin P = \sin Q$.

How then, can we prove the fact above?

We can use another trig-identity to produce the proof.

And the other identity is: $\sin(a \pm b) = \sin a \cos b \pm \cos a \sin b$.

And we have: **sin π = 0** and **cos π = -1**.

So we get: **sin (π − θ) = sin π cos θ − cos π sin θ = 0·cos θ − (-1)·sin θ = sin θ.**

And we can do the proof graphically, too.

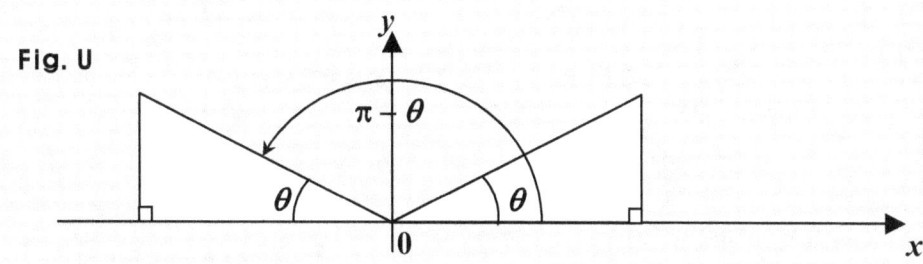

Fig. U

So to begin with, we can see that in each triangle, the opposite is the same.

And next, using the triangle on the left, we get: **sin (π − θ).**

And also, using the triangle on the right, we get: **sin θ.**

We know the sine is the ratio of the opposite to the hypotenuse, that is, the opposite over the hypotenuse. And both triangles have the same hypotenuse.

So we get: **sin (π − θ) = sin θ.**

In other words, if *A* + *B* = π, we get: **sin *A* = sin *B*.**

And using the sine rule, we can come up with an expression showing the ratio between the three sides in terms of the sines of all the three angles.

That is to say that:

We have: $\dfrac{a}{\sin A} = \dfrac{b}{\sin B} = \dfrac{c}{\sin C} = 2R.$ So we get: $a : b : c = \sin A : \sin B : \sin C.$

And of course, we can get these, too:

$$a : b = \sin A : \sin B, \quad b : c = \sin B : \sin C, \quad \text{and } a : c = \sin A : \sin C.$$

How come though?

First, we get: $\dfrac{a}{\sin A} = \dfrac{b}{\sin B} = \dfrac{c}{\sin C} = 2R \Rightarrow \dfrac{a}{\sin A} = 2R \Rightarrow a = 2R \sin A.$

And by the same token, we can get:

$$\dfrac{b}{\sin B} = 2R \Rightarrow b = 2R \sin B, \text{ and } \dfrac{c}{\sin C} = 2R \Rightarrow c = 2R \sin C.$$

So we get: $a : b : c = 2R \sin A : 2R \sin B : 2R \sin C = \sin A : \sin B : \sin C.$

And thus, we get: $a : b : c = \sin A : \sin B : \sin C.$

So we can say that:

In a triangle, the ratio between the sides is the same as the ratio between the *sines* of the angles facing the sides.

Examples in The Sine Rule

Doing all these examples, refer to the triangle ABC below:

Fig. 0

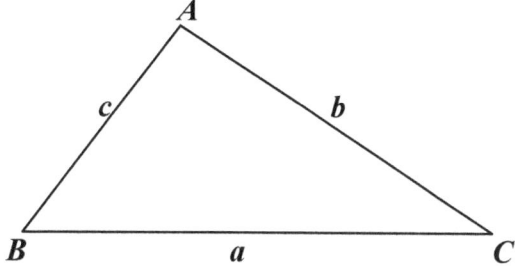

0.   Show that:

$\sin A + \sin B > \sin C, \sin B + \sin C > \sin A,$ **and** $\sin C + \sin A > \sin B$.

1.   Assuming $2b = a + c$, show the relation among the angles A, B and C.

2.   Assuming $A : B : C = 3 : 4 : 5$, find $a : b : c$.

3.   The equation below is for x, is quadratic, and has two identical solutions, that is, a double root. What kind of triangle then, is the triangle ABC?

$(\sin C + \cos A)x^2 + (2\cos B)x - (\sin C - \cos A) = 0.$

4.   Assuming: $a = 100$, $B = \pi/3$, and $C = 5\pi/12$, find A, b, and c.

5.   Assuming: $b = 15$, $c = 15\sqrt{3}$, and $B = \pi/6$, find A, a, and C.

Suggestions or Solutions
To the **Problem** in the Example **0**

Show that: sin *A* + sin *B* > sin *C*, sin *B* + sin *C* > sin *A*, and sin *C* + sin *A* > sin *B* for the triangle *ABC* below:

Fig. 0.0

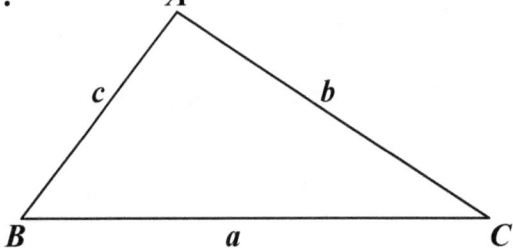

We have: $\dfrac{a}{\sin A} = \dfrac{b}{\sin B} = \dfrac{c}{\sin C} = 2R$, $a + b > c$, $b + c > a$, and $c + a > b$.

So first, we get: $a = 2R \sin A$, $b = 2R \sin B$, and $c = 2R \sin C$.

Thus, next, we get:

$a + b > c \Rightarrow 2R \sin A + 2R \sin B > 2R \sin C \Rightarrow \sin A + \sin B > \sin C.$

$b + c > a \Rightarrow 2R \sin B + 2R \sin C > 2R \sin A \Rightarrow \sin B + \sin C > \sin A.$

$c + a > b \Rightarrow 2R \sin C + 2R \sin A > 2R \sin B \Rightarrow \sin C + \sin A > \sin B.$

If not quite sure of the idea behind the processes above, follow the steps below:

Each of the relations given looks like a property in a triangle. What property?

It is called Triangle Inequality, and says that the sum of two sides is grater than the other side, in a triangle, of course. So for instance, we get: $a + b > c$, $b + c > a$, and $c + a > b$.

How then, can use the fact, the inequality above?

Looking at the relations given, we can notice that they are made of the sines. So we may want to try the sine rule. What then, is the sine rule?

Referring to the triangle above, we can put it this way: $\dfrac{a}{\sin A} = \dfrac{b}{\sin B} = \dfrac{c}{\sin C} = 2R.$

What then, is **R**?

It is the radius of a circle called a circumcircle, which passes through all the vertices of the triangle **ABC**, which is thus, circumscribed by the circle of radius **R**.

So putting the triangle **ABC** in the circle, we can put it the way below:

Fig. 0.1

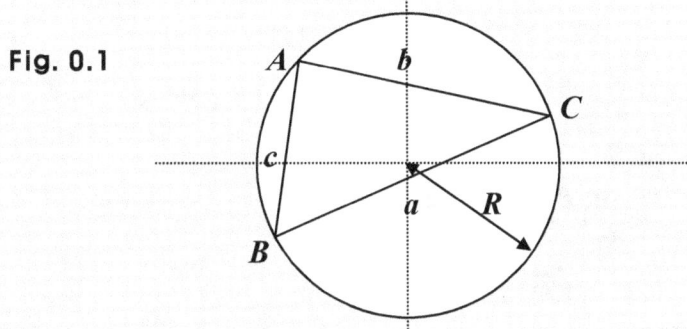

How then, can we use the sine rule?

We can extract the three sides from the sine rule, and set up the inequality.

And the sine rule is: $\dfrac{a}{\sin A} = \dfrac{b}{\sin B} = \dfrac{c}{\sin C} = 2R.$

So we get: $a = 2R \sin A$, $b = 2R \sin B$, and $c = 2R \sin C$.

And setting up the inequality, we get: $a + b > c \Rightarrow 2R \sin A + 2R \sin B > 2R \sin C.$

We know **2R** is positive. So dividing by **2R** both sides, we get: $\sin A + \sin B > \sin C.$

And the same is true, too, for the other two relations.

So setting up again, the inequality, we get: $b + c > a \Rightarrow 2R \sin B + 2R \sin C > 2R \sin A$.

And dividing by $2R$ both sides, we get: $\sin B + \sin C > \sin A$.

Setting up again, the inequality, we get: $c + a > b \Rightarrow 2R \sin C + 2R \sin A > 2R \sin B$.

And dividing again, by $2R$ both sides, we get: $\sin C + \sin A > \sin B$.

In short:

We have: $\dfrac{a}{\sin A} = \dfrac{b}{\sin B} = \dfrac{c}{\sin C} = 2R$, $a + b > c$, $b + c > a$, and $c + a > b$.

So first, we get: $a = 2R \sin A$, $b = 2R \sin B$, and $c = 2R \sin C$.

Thus, next, we get:

$a + b > c \Rightarrow 2R \sin A + 2R \sin B > 2R \sin C \Rightarrow \sin A + \sin B > \sin C$.

$b + c > a \Rightarrow 2R \sin B + 2R \sin C > 2R \sin A \Rightarrow \sin B + \sin C > \sin A$.

$c + a > b \Rightarrow 2R \sin C + 2R \sin A > 2R \sin B \Rightarrow \sin C + \sin A > \sin B$.

Suggestions or Solutions
To the **Problem** in the Example 1

Assuming $2b = a + c$, show the relation among the angles A, B and C for the triangle below:

Fig. 1.0

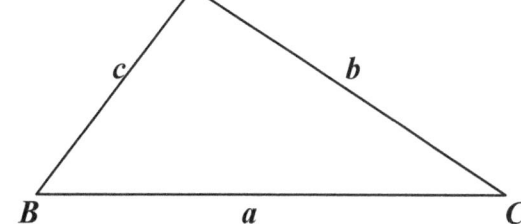

What do we mean by showing the relation among the angles A, B and C?

We can show it producing a math expression in terms of A, B and C.

We know that A, B and C are angles. How then, can we relate angles?

Using an angle, we can express a ratio. What ratio, though?

It is a trig-ratio as the sine, the cosine, and the tangent.

So we can try using such a ratio setting up the relation among the angles A, B and C.

We cannot however, just set up a relation. The relation has to be on a specific basis. What can be the basis then?

We are given a relation between the sides in the triangle that has the angles. And the relation is the expression where $2b = a + c$.   So we want to use it.   How though?

We have the sine rule, and we can extract the three sides from the sine rule, and set up the relation. What then, is the sine rule?

The sine rule is: $\dfrac{a}{\sin A} = \dfrac{b}{\sin B} = \dfrac{c}{\sin C} = 2R.$

So we get: $a = 2R\sin A$, $b = 2R\sin B$, and $c = 2R\sin C$.

And setting up the relation, we get: $2b = a + c \Rightarrow 2 \cdot 2R\sin B = 2R\sin A + 2R\sin C.$

We know $2R$ is positive. So dividing by $2R$ both sides, we get: $2\sin B = \sin A + \sin C.$

Suggestions or Solutions
To the **Problem** in the Example **2**

Find $a : b : c$ **assuming** $A : B : C = 3 : 4 : 5$ **for the triangle** ABC **below:**

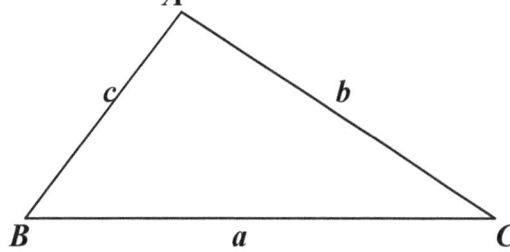

Fig. 2.0

We want to find the ratio between the sides in a triangle, based on the ratio between the angles in the triangle.

So it seems we want to make use of the ratio between the angles. How?

We can put the ratio in terms of the angles. How?

We have the sine rule where: $\dfrac{a}{\sin A} = \dfrac{b}{\sin B} = \dfrac{c}{\sin C} = 2R.$

So we can get: $a = 2R \sin A$, $b = 2R \sin B$, and $c = 2R \sin C$.

And thus, we can for now, put the ratio the way below:

$a : b : c = 2R \sin A : 2R \sin B : 2R \sin C = \sin A : \sin B : \sin C.$

So we now have: $a : b : c = \sin A : \sin B : \sin C$. What then, is the next?

We have not use the fact given, and the fact is the ratio between the angles. So we want to make use of the ratio, which is: $A : B : C = 3 : 4 : 5$. How then, can we use it?

We have another property in a triangle. And the property is the fact that the sum of the three angles in a triangle is $180°$, which is π.

So using the fact above, we can actually get the values of the angles, **A**, **B**, and **C**.

And finding them, we get: $A = \pi\frac{3}{3+4+5} = \pi\frac{3}{12} = \pi/4$, $B = \pi\frac{4}{12} = \pi/3$, and $C = 5\pi/12$.

So we get: **a : b : c = sin π/4 : sin π/3 : sin 5π/12**.

And we know: $\sin \pi/4 = \frac{\sqrt{2}}{2}$, and $\sin \pi/3 = \frac{\sqrt{3}}{2}$. What about **sin 5π/12** though?

We know $5\pi/12$ is the governing angle.

So using a right triangle where an angle is $5\pi/12$, which is $75°$, we should be able to get the value of the trig-ratio, **sin 75°**. And thus, forming such a triangle, we can get:

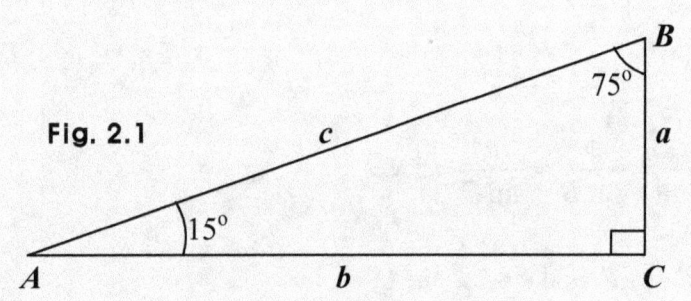

Fig. 2.1

We know that the sine is: the opposite over the hypotenuse.

So we can set: **sin 75° = b/c**.

And in fact, we found the ratio **b/c** in the example 0 in the section, **Triangles and Trigonometry**.

And the ratio is: $\frac{\sqrt{2}+\sqrt{6}}{4}$. So we get: $\sin 75° = \frac{\sqrt{2}+\sqrt{6}}{4}$.

And thus, we now can set: $a : b : c = \frac{\sqrt{2}}{2} : \frac{\sqrt{3}}{2} : \frac{\sqrt{2}+\sqrt{6}}{4}$, which is: $2\sqrt{2} : 2\sqrt{3} : \sqrt{2}+\sqrt{6}$.

So we get: $a : b : c = 2\sqrt{2} : 2\sqrt{3} : \sqrt{2}+\sqrt{6}$.

Suggestions or Solutions
To the **Problem** in the Example **3**

The equation below is for *x*, is quadratic, and has two identical solutions, that is, a double root. What kind of triangle then, is the triangle *ABC*?

$(\sin C + \cos A)x^2 + (2\cos B)x - (\sin C - \cos A) = 0.$

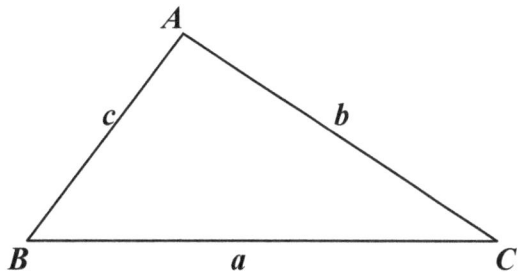

Fig. 3.0

Assuming ***D*** is the discriminant, we get:

$D = (2\cos B)^2 - 4(\sin C + \cos A)\{-(\sin C - \cos A)\} = 0$

$\Rightarrow 4\cos^2 B + 4(\sin C + \cos A)(\sin C - \cos A) = 0.$

$\Rightarrow \cos^2 B + (\sin C + \cos A)(\sin C - \cos A) = 0 \Rightarrow \cos^2 B + \sin^2 C - \cos^2 A = 0.$

And we have a trig-identity where $\sin^2 \theta + \cos^2 \theta = 1$. So we get:

$\sin^2 A + \cos^2 A = 1 \Rightarrow \cos^2 A = 1 - \sin^2 A$, and

$\sin^2 B + \cos^2 B = 1 \Rightarrow \cos^2 B = 1 - \sin^2 B$. Thus, we get:

$\cos^2 B + \sin^2 C - \cos^2 A = 1 - \sin^2 B + \sin^2 C - (1 - \sin^2 A) \Rightarrow \sin^2 A + \sin^2 C = \sin^2 B.$

And also, we have: $\dfrac{a}{\sin A} = \dfrac{b}{\sin B} = \dfrac{c}{\sin C} = 2R.$

So we get: $\sin A = a/2R$, $\sin B = b/2R$, and $\sin C = c/2R$.

Thus, we get: $\sin^2 A + \sin^2 C = \sin^2 B \Rightarrow (a/2R)^2 + (c/2R)^2 = (b/2R)^2 \Rightarrow a^2 + c^2 = b^2.$

So the triangle ***ABC*** is a right triangle where ***b*** is the hypotenuse.

If not quite sure of the idea behind the processes above, follow the steps below:

If a quadratic equation $ux^2 + vx + w = 0$ has a double root, we get: $D = 0$, where D is called the discriminant. And D is in this case, $v^2 - 4uw$.

So since the equation given has a double root, applying the discriminant D to the equation, we get: $D = (2\cos B)^2 - 4(\sin C + \cos A)\{-(\sin C - \cos A)\} = 0$.

That is, we get: $4\cos^2 B + 4(\sin C + \cos A)(\sin C - \cos A) = 0$.

And dividing by 4 both sides, we get: $\cos^2 B + (\sin C + \cos A)(\sin C - \cos A) = 0$.

And simplifying it, we get: $\cos^2 B + \sin^2 C - \cos^2 A = 0$.

How then, do we know the kind the triangle ABC belongs to?

Finding the relation among the sides, a, b, and c, or knowing about the angles, A, B, and C, we can see the kind. In this case, we can find the relation among the sides easier. How though?

We can use two facts, one is a trig-identity, and the other is the sine rule. What trig-identity?

It is a trig-identity where $\sin^2 \theta + \cos^2 \theta = 1$. So we get:

$\sin^2 A + \cos^2 A = 1 \Rightarrow \cos^2 A = 1 - \sin^2 A$, and

$\sin^2 B + \cos^2 B = 1 \Rightarrow \cos^2 B = 1 - \sin^2 B$. Thus, we get:

$\cos^2 B + \sin^2 C - \cos^2 A = 1 - \sin^2 B + \sin^2 C - (1 - \sin^2 A) \Rightarrow \sin^2 A + \sin^2 C = \sin^2 B$.

And moving on to the sine rule, we have: $\dfrac{a}{\sin A} = \dfrac{b}{\sin B} = \dfrac{c}{\sin C} = 2R$.

So we get: $\sin A = a/2R$, $\sin B = b/2R$, and $\sin C = c/2R$.

And thus, we get: $\sin^2 A + \sin^2 C = \sin^2 B \Rightarrow (a/2R)^2 + (c/2R)^2 = (b/2R)^2$.

And multiplying by $4R^2$ both sides, we get: $a^2 + c^2 = b^2$, which indicates a right triangle where b is the hypotenuse.

So the triangle ABC is a right triangle where b is the hypotenuse, and B is a right angle, since the hypotenuse faces the right angle.

In short:

Assuming D is the discriminant, we get:

$D = (2\cos B)^2 - 4(\sin C + \cos A)\{-(\sin C - \cos A)\} = 0$

$\Rightarrow 4\cos^2 B + 4(\sin C + \cos A)(\sin C - \cos A) = 0$.

$\Rightarrow \cos^2 B + (\sin C + \cos A)(\sin C - \cos A) = 0 \Rightarrow \cos^2 B + \sin^2 C - \cos^2 A = 0$.

And we have a trig-identity where $\sin^2 \theta + \cos^2 \theta = 1$. So we get:

$\sin^2 A + \cos^2 A = 1 \Rightarrow \cos^2 A = 1 - \sin^2 A$, and

$\sin^2 B + \cos^2 B = 1 \Rightarrow \cos^2 B = 1 - \sin^2 B$. Thus, we get:

$\cos^2 B + \sin^2 C - \cos^2 A = 1 - \sin^2 B + \sin^2 C - (1 - \sin^2 A) \Rightarrow \sin^2 A + \sin^2 C = \sin^2 B$.

And also, we have: $\dfrac{a}{\sin A} = \dfrac{b}{\sin B} = \dfrac{c}{\sin C} = 2R$.

So we get: $\sin A = a/2R$, $\sin B = b/2R$, and $\sin C = c/2R$.

Thus, we get: $\sin^2 A + \sin^2 C = \sin^2 B \Rightarrow (a/2R)^2 + (c/2R)^2 = (b/2R)^2 \Rightarrow a^2 + c^2 = b^2$.

So the triangle ABC is a right triangle where b is the hypotenuse.

Suggestions or Solutions
To the **Problem** in the Example **4**

Find A, b, and c assuming: $a = 100$, $B = \pi/3$, and $C = 5\pi/12$ for the triangle below:

Fig. 4.0

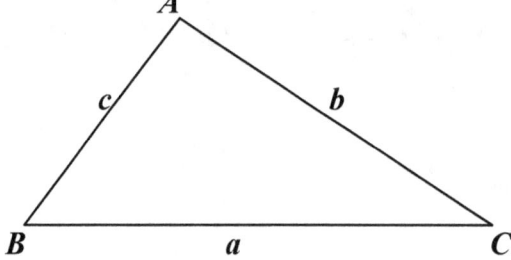

To begin with, we know two angles in a triangle.

So we can get the other angle, since the sum of the three angles in a triangle is π.

And the two are **$B = \pi/3$, and $C = 5\pi/12$.**

Thus, we get: $A = \pi - B - C = \pi - \pi/3 - 5\pi/12 = \pi - 4\pi/12 - 5\pi/12 = \pi - 9\pi/12 = 3\pi/12$.

So we get: $A = \pi/4$. What then, is the next?

Next, we have the sine rule: $\dfrac{a}{\sin A} = \dfrac{b}{\sin B} = \dfrac{c}{\sin C} = 2R$.

So we get: $\dfrac{100}{\sin \pi/4} = \dfrac{b}{\sin \pi/3} = \dfrac{c}{\sin 5\pi/12} = 2R$. What then?

We know: $\sin \pi/4 = \frac{\sqrt{2}}{2}$, $\sin \pi/3 = \frac{\sqrt{3}}{2}$, and $\sin 5\pi/12 = \frac{\sqrt{2}+\sqrt{6}}{4}$, from the example 2.

So first, we can get: $\dfrac{100}{\sin \pi/4} = \dfrac{b}{\sin \pi/3}$

$\Rightarrow b = \dfrac{100\sin \pi/3}{\sin \pi/4} = \dfrac{100\frac{\sqrt{3}}{2}}{\frac{\sqrt{2}}{2}} = \dfrac{100\sqrt{3}}{\sqrt{2}} = \dfrac{100\sqrt{6}}{2} = 50\sqrt{6} \Rightarrow b = 50\sqrt{6}$.

And next, we can get: $\dfrac{100}{\sin \pi/4} = \dfrac{c}{\sin 5\pi/12}$

$$\Rightarrow c = \frac{100\sin 5\pi/12}{\sin \pi/4} = \frac{100\frac{\sqrt{2}+\sqrt{6}}{4}}{\frac{\sqrt{2}}{2}} = \frac{50(\sqrt{2}+\sqrt{6})}{\sqrt{2}} = \frac{50\sqrt{2}(\sqrt{2}+\sqrt{6})}{2}$$

$$= 25(2+2\sqrt{3}) = 50(1+\sqrt{3}) \;\Rightarrow c = 50(1+\sqrt{3}).$$

Suggestions or Solutions
To the **Problem** in the Example **5**

Find A, a, and C assuming: $b = 15$, $c = 15\sqrt{3}$, and $B = \pi/6$ for the triangle below:

Fig. 5.0

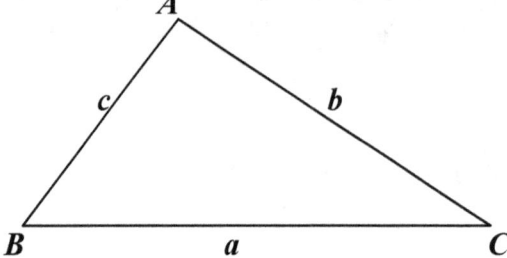

To begin with, we have the sine rule: $\dfrac{a}{\sin A} = \dfrac{b}{\sin B} = \dfrac{c}{\sin C} = 2R$.

So we get: $\dfrac{a}{\sin A} = \dfrac{15}{\sin \pi/6} = \dfrac{15\sqrt{3}}{\sin C} = 2R$. And we know: $\sin \pi/6 = 1/2$. So?

So first, we can get: $\dfrac{15}{\sin \pi/6} = \dfrac{15\sqrt{3}}{\sin C} \Rightarrow 15 \sin C = 15\sqrt{3} \sin \pi/6 = \frac{15\sqrt{3}}{2} \Rightarrow \sin C = \frac{\sqrt{3}}{2}$.

Thus, we get: $C = \pi/3$ or $\pi - \pi/3 = 2\pi/3$ since we have: $\sin (\pi - \theta) = \sin \theta$ for $0 \le \theta \le \pi/2$.
And we know the sum of the three angles in a triangle is π. So?

So assuming first, $C = \pi/3$, we get: $A + B + C = A + \pi/6 + \pi/3 = \pi \Rightarrow A = \pi/2$.

So next, we can get: $\dfrac{a}{\sin \pi/2} = \dfrac{15}{\sin \pi/6} = \dfrac{15}{\frac{1}{2}} = 30 \Rightarrow a = 30 \sin \pi/2 = 30$, for $\sin \pi/2 = 1$.

And assuming next, $C = 2\pi/3$, we get: $A + B + C = A + \pi/6 + 2\pi/3 = \pi \Rightarrow A = \pi/6$.

And next, we can get: $\dfrac{a}{\sin \pi/6} = \dfrac{15}{\sin \pi/6} \Rightarrow a = 15$.

And thus, we get: $A = \pi/2$, $a = 30$, and $C = \pi/3$. $A = \pi/6$, $a = 15$, and $C = 2\pi/3$.

6. The Cosine Rule

The cosine is a trig-ratio, and the cosine of an angle is the ratio of the adjacent to the hypotenuse in a right triangle, where the hypotenuse and the adjacent makes the angle.

So in short, the cosine is the adjacent over the hypotenuse in a right triangle.

What then, is the cosine rule?

It is often called the law of cosines, too, and we can call it the cosine formula, also.

And as stated above, we get the trig-ratio called the cosine from a right triangle. Using the cosine rule though, we can apply it to any triangle, and thus, not a right triangle only.

And by means of the cosine rule, we can express one side in a triangle T, for instance, in terms of the other two sides and the cosine of the angle facing the one side.

So for instance, assuming the three angles are A, B, and C, the three sides are a, b, and c, and expressing the side a using the cosine rule, we can put it the way below:

Fig. 0

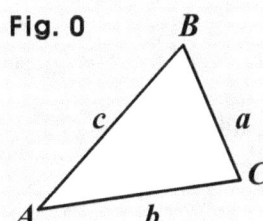

$c^2 = a^2 + b^2 - 2ab \cos C.$

And the same is true for the other sides, too:

$a^2 = b^2 + c^2 - 2bc \cos A.$

$b^2 = c^2 + a^2 - 2ca \cos B.$

And assuming in particular, the facing angle C is 90°, we get: $c^2 = a^2 + b^2$.

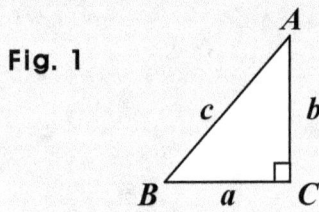

Fig. 1

And of course, the same is true for the other angles, too.

So we get: $a^2 = b^2 + c^2$ if A is $\pi/2$, and $b^2 = c^2 + a^2$ if B is 90°.

And the proof is rather easier than the one for the sine rule.

As in the case of the sine rule though, in the cosine rule, too, one of the three angles can be obtuse as well as acute and 90°.

That is, we can have either of three cases as follows: $0 < A < \pi$, $0 < B < \pi$, and $0 < C < \pi$.

So to begin with, assuming T is an *acute* triangle, we can put it the way below:

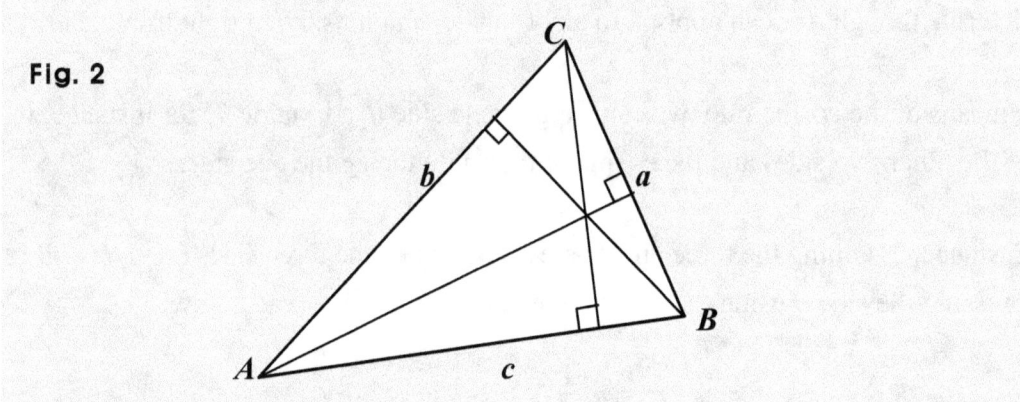

Fig. 2

And we know multiplying the hypotenuse by the cosine, we get the adjacent.

That is, **hypotenuse $\cdot$ cos θ = hypotenuse $\cdot \dfrac{\text{adjacent}}{\text{hypotenuse}}$**.

So first, we get:

$a = b \cos C + c \cos B$ .... **(0)**

$b = c \cos A + a \cos C$ .... **(1)**

$c = a \cos B + b \cos A$ .... **(2)**

And next:

Multiplying **(0)** by a, we get: $a^2 = ab \cos C + ac \cos B$ .... **(3)**

Multiplying **(1)** by b, we get: $b^2 = bc \cos A + ab \cos C$ .... **(4)**

Multiplying **(2)** by c, we get: $c^2 = ac \cos B + bc \cos A$ .... **(5)**

And next, adding together **(4)** and **(5)**, we get:

$b^2 + c^2 = bc \cos A + ab \cos C + ac \cos B + bc \cos A = ab \cos C + ac \cos B + 2bc \cos A.$

And we have: $a^2 = ab \cos C + ac \cos B$ .... **(3)**.

So we get: $b^2 + c^2 = a^2 + 2bc \cos A$. And thus, we get: $a^2 = b^2 + c^2 - 2bc \cos A.$

And next, adding together **(5)** and **(3)**, we get:

$c^2 + a^2 = ac \cos B + bc \cos A + ab \cos C + ac \cos B = ab \cos C + bc \cos A + 2ac \cos B.$

And we have: $b^2 = bc \cos A + ab \cos C$ .... **(4)**.

So we get: $c^2 + a^2 = b^2 + 2ac \cos B$. And thus, we get: $b^2 = c^2 + a^2 - 2ca \cos B.$

And next, adding together **(3)** and **(4)**, we get:

$a^2 + b^2 = ab \cos C + ac \cos B + bc \cos A + ab \cos C = ac \cos B + bc \cos A + 2ab \cos C.$

And we have: $c^2 = ac \cos B + bc \cos A$ .... **(5)**.

So we get: $a^2 + b^2 = c^2 + 2ab \cos C$. And thus, we get: $c^2 = a^2 + b^2 - 2ab \cos C.$

And let's next, move on to the case where the triangle **T** is an *obtuse* triangle.
Then first, we can put it the way below:

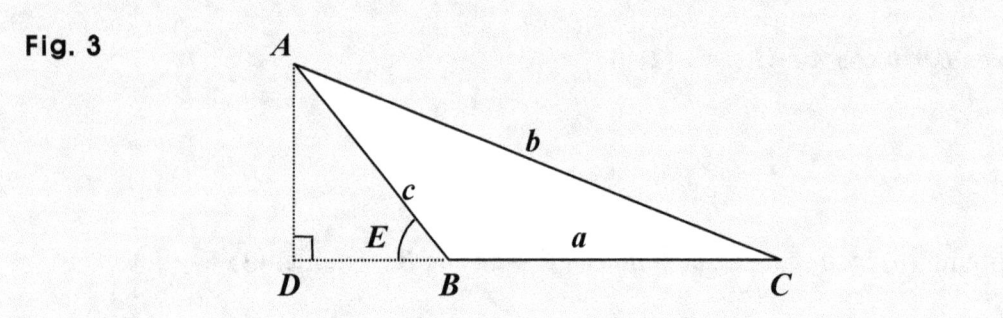

Fig. 3

Then first, we get: $a = \overline{CD} - \overline{BD} = b \cos C - c \cos E = b \cos C - c \cos (\pi - B)$.

And we have a trig-identity where $\cos (\pi - \theta) = -\cos \theta$.   How come?

We can see how it is the case graphically:

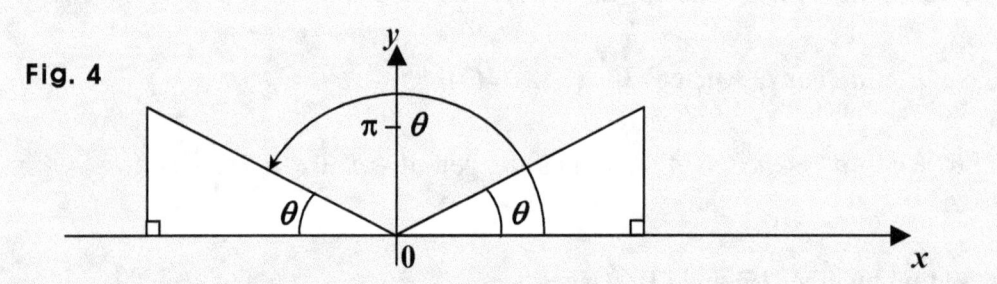

Fig. 4

The right triangle on the right is normal, but the one on the left is not, and can be said to be *transcendental*. In a right triangle transcendental, either or both of the lengths of the opposite and the adjacent can be negative or 0 as well as positive.

So the right triangle on the left is transcendental, and its adjacent is negative.

The magnitude of the adjacent is though, the same as that of the adjacent in the right triangle on the right. And no matter what right triangle we may work with, taking the cosine, we still get the ratio of the adjacent to the hypotenuse in the right triangle.

So the cosine of an angle $\pi - \theta$ is the ratio of the adjacent to the hypotenuse in the right triangle on the left, but the adjacent is negative. So we use $-\cos \theta$ as $\cos (\pi - \theta)$.
How come?

That's because taking the **cos θ** in the right triangle on the right, we get the ratio of the *adjacent* to the *hypotenuse* where the *adjacent* is positive.

Note that no matter what right triangle it may be, the hypotenuse is always positive.

And we have these, too: **sin (π − θ) = sin θ**, and **tan (π − θ) = −tan θ**.

So we get: $a = b \cos C - c \cos (\pi - B) = b \cos C + c \cos B$.

And thus, we get: $a = b \cos C + c \cos B$.

And moving next, on to the case of **c**, we can begin with putting the triangle the way below:

Fig. 5

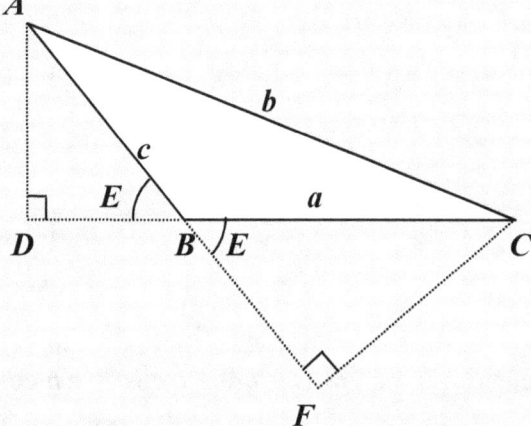

Then first, we get: $c = \overline{AF} - \overline{BF} = b \cos A - a \cos E = b \cos A - a \cos (\pi - B)$.

And we have a trig-identity where **cos (π − θ) = −cos θ**.

So we get: $c = b \cos A - a \cos (\pi - B) = b \cos A + a \cos B$.

And thus, we get: $c = b \cos A + a \cos B$.

And moving next, on to the case of **b**, we can begin with putting the triangle the way below:

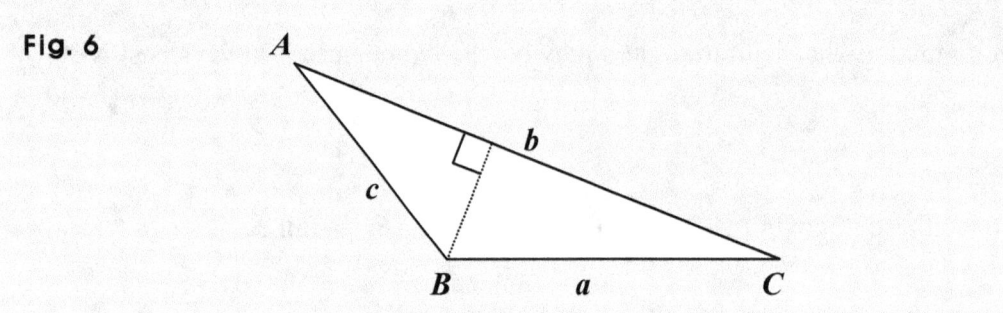

Fig. 6

Then, we can simply get: $b = c \cos A + a \cos C$.

And doing the same in the case where T is the acute triangle, we can get:
$a^2 = b^2 + c^2 - 2bc \cos A$, $b^2 = c^2 + a^2 - 2ca \cos B$, and $c^2 = a^2 + b^2 - 2ab \cos C$.

And let's next, move on to the case where the triangle T is an obtuse triangle. Then first, we can put it the way below:

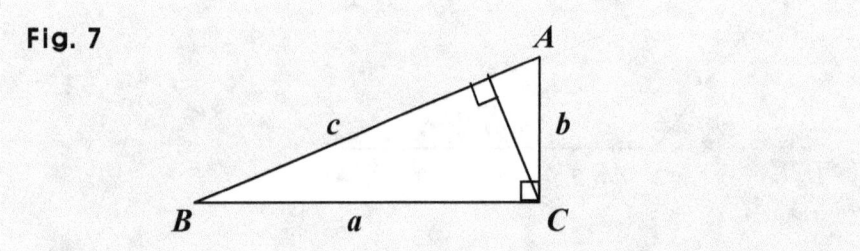

Fig. 7

Then first, we get: $a = c \cos B = 0 + c \cos B = b \cdot 0 + c \cos B = b \cos C + c \cos B$, because $C = \pi/2$, and $\cos \pi/2 = 0$. So we get; $a = b \cos C + c \cos B$.

And next, we get: $b = c \cos A = c \cos B + 0 = c \cos A + a \cdot 0 = c \cos A + a \cos C$, because $C = \pi/2$, and $\cos \pi/2 = 0$. So we get; $b = c \cos A + a \cos C$.

And we can simply get: $c = a \cos B + b \cos A$.

And doing the same in the case where T is the acute triangle, we can get:
$a^2 = b^2 + c^2 - 2bc \cos A$, $b^2 = c^2 + a^2 - 2ca \cos B$, and $c^2 = a^2 + b^2 - 2ab \cos C$.

www.ingramcontent.com/pod-product-compliance
Lightning Source LLC
Chambersburg PA
CBHW081449170526
45166CB00008B/2372